An Enduring Quest

The Story of Purdue Industrial Engineers

AN ENDURING QUEST

The Story of Purdue
Industrial Engineers

Ferd Leimkuhler

Purdue University Press
West Lafayette, Indiana

The images on pages 39, 52-58, 66, 68, 70, 84-87, 96, and 161 are printed courtesy of Purdue University Libraries, Archives, and Special Collections. Other images are printed courtesy of the Purdue University School of Industrial Engineering on pages 112-115 and pages 215-223, J. T. Black on the cover and pages 82 and 86, SCRAN Ltd. on page 25, James Solberg on page 127 and 128, Natalie Leimkuhler on page 155, Takao Mundel on page 165, and Anne Pritsker on page 195. The centerpiece of color photographs are printed courtesy of Patrick Whalen, the Detroit Institute of Art, and the Indiana University Archives. The cover design is by Natalie Powell.

Library of Congress Cataloging-in-Publication Data

Leimkuhler, Ferdinand F.
  An enduring quest : the story of Purdue industrial engineers / Ferd Leimkuhler.
      p. cm.
  Includes bibliographical references.
  ISBN 978-1-55753-544-3
1. Industrial engineering--United States. 2. Industrial engineers--United States. 3. Purdue University. I. Purdue University. II. Title.

  T55.7.L54 2009
  670.973--dc22
                                2009002270

CONTENTS

Author's Preface

While aimed at engineers and students interested in the field, the author hopes this book will help others understand how the industrialization of America came about and helped changed the world. Because of its profound effect on our everyday life, this technology is much too important to be left to engineers alone. Everyone needs to participate in a free and open discussion of how technology is used, and I hope this book will contribute to such a conversation.

Writing the book was a way of saying thanks to my many students and colleagues, and to my teachers, Howard Ellis, Rob Roy, George Hawkins, Philip Morse, and Moshe Barash. I am very grateful to people at Purdue who helped in this effort, especially Charlotte Erdmann, Dianna Gilroy, Dan Folta, Sammie Morris, and Nagabhushana Prabhu. Special thanks go to my wife, Natalie, for her encouragement and patience, daughters Meg and Jeanne, and above all to my son Tom who was chief critic, editor, and muse.

Ferd Leimkuhler
Berkeley, 2009

Foreword by Steven C. Beering

It was nearly forty years ago when I first met Sam Regenstrief. He was an enterprising small-town manufacturer of dishwashers who proposed that we approach the management of health care in the same way that he produced the working parts of dishwashers. On the surface, his suggestion seemed preposterous. He was persistent, however, and even offered to fund a building to house an outpatient clinic and a center for healthcare research at the Indiana University School of Medicine. In due course we assembled a number of very bright young Purdue industrial engineers who, together with a cadre of committed internal medicine professors, created the Regenstrief Institute and Clinic and revolutionized our handling of outpatients and their records. As a classically educated physician and medical school dean, I knew about individual patient care, but had not appreciated the powerful influence of industrial engineering. Today we employ interdisciplinary approaches in medicine and basic science and we are proud of the contributions of systems research to hospital management and health care delivery.

Dr. Ferd Leimkuhler, pioneering professor of industrial engineering, has chronicled the remarkable history of this discipline at Purdue University and in America. His encyclopedic account is fascinating both in breadth and detail. He carefully describes the profound contributions of industrial engineering to the resolution of many major societal problems over the past fifty years. We can now earnestly hope that the lessons learned will be applied to addressing global crises in the environment, energy, and the world economy. Fundamental to our collective future welfare will be a re-emphasis of education at all levels. We have the knowledge, the skills, and the means to succeed.

Steven C. Beering, President Emeritus, Purdue University

Foreword by Gerald Nadler

I was introduced to Purdue in July sixty-five years ago, in 1943. The US Navy V-12 program assigned me there from the mechanical engineering (ME) program I had started at the University of Cincinnati. Little did I know how that transfer would so dramatically affect my life. The ME bachelor's degree I received two years later would have been rather uneventful if it hadn't given me the opportunity to take Marvin Mundel's motion and time study course as an elective in my last semester. It opened my eyes to the wonderful perspectives of industrial engineering and led me to pursue graduate studies in that burgeoning field.

The early maturation period for the IE profession was a most exciting time to enter the IE graduate program at Purdue. My fellow graduate students at one time or another (Janet Armstrong, Irving Lazarus, Robert Lehrer, Donald Malcolm, and Harold Smalley, to name a few I remember) were equally engaged in assuring that the profession would become a major program at Purdue and in the engineering world at large.

The faculty members (Harold Amrine, Robert Fields, Lillian Gilbreth, Marvin Mundel, Halsey Owen, and Wally Richardson) also made us feel we were pioneers in the emerging formalization of industrial engineering educational programs as a foundation for the IE profession. Lillian Gilbreth was a great inspiration to all of us as we discussed what IE should be. We were early enthusiasts when the American Institute of Industrial Engineers was founded in 1948. That Purdue set up its IE program a few years after I and my fellow graduate students moved on made me proud that we had perhaps helped put a foundation under what has now become a premier school of industrial engineering.

I watched with pleasure as Harold Amrine led Purdue's IE program through adolescence and was especially delighted when Ferd Leimkuhler's long period of leadership brought the program to its outstanding adulthood. He is to be highly complimented on the history he has prepared. It is particularly valuable because of his explanation of the larger societal context that led to the profession of IE as well as its trajectory at Purdue. Anyone in the field in addition to those with some association with Purdue will benefit from the insights in this book.

Gerald Nadler

1 | An Enduring Quest

The political upheaval of the American Revolution in 1776 coincided with a wave of manufacturing innovation in Great Britain that crossed the ocean and became a major factor in the growth of the new nation. In the two centuries after its founding, America went from being an essentially agrarian society to being an unrivaled industrial powerhouse. At the forefront of this change were industrial engineers who created the large-scale production and service systems that are basic to modern society.

Today a new era of innovation is underway that is transforming American industry, making it more flexible, decentralized, and knowledge intensive. The present dependence on materials, energy, labor, and capital is giving way to the development of new ways to exploit information and knowledge resources. Industrial engineers are helping to meet this challenge through their experience in designing large-scale systems for global production of goods and services and their unique interest in the interaction of humans and technology.

This book describes how industrial engineering evolved over the past two centuries by telling the story of its growth at Purdue University, where it helped shape the University itself and made essential contributions to the field. The term "industrial engineering" came into popular use around 1900 and was first used to describe a course at Purdue in 1908, a professorship in 1919, graduate degrees in 1937, and bachelor's degrees in 1955. When first taught in 1879, it was called "practical mechanics," and later, "general engineering," before these programs were merged into the School of Industrial Engineering we know today.

The events that led up to Purdue opening its doors in 1874 started a century earlier in Great Britain with a manufacturing revolution that began at the time of the American Revolution. In 1776, Adam Smith made an epic study, *An Inquiry into the Nature and Causes of* the Wealth of Nations, in which he said that the increase in manufacturing productivity in factories was due to a combination of new work methods and machine tools. Two outstanding inventors at the time were Henry Maudslay in England and Eli Whitney in America. Maudslay's amazingly accurate lathe became the model for developing the English system of precision manufacturing that enabled Britain to excel in bringing steam power to factories, railroads, and ships.

Whitney invented the cotton gin and a multitooled milling machine, but his greater accomplishment was conceiving the American system of mass production in which the work of a master craftsman was divided into a sequence of tasks so that unskilled workmen using special tools could make an acceptable replica of a masterwork. With mass production and his earlier invention of the cotton gin, which enabled the South to sell cotton to England, Whitney almost single-handedly revived the American economy after the Revolutionary War and set the course of its industrial development through the Civil War and into the twentieth century.

Mass production in factories was started in 1771 by the entrepreneur Richard Arkwright who began spinning cotton in a carefully managed mill. He was the first to use James Watt's steam engine in a factory setting and planned the New Lanark Mill, Britain's largest, later owned by Robert Owen. The factory system had a social result that was much more significant than the outpouring of consumer goods. It broke the grip of the craft guilds over manufacturing and prompted an exodus of workers from farms to factories. Factories became, in effect, public schools to teach trades to young people who wanted to escape medieval farm life even at the high cost of uncertain employment and poor working and living conditions in factory towns.

It later came to be seen that mastering the new factory technology was a way to teach "engineering design" to beginning students in the new public technical schools created after the Civil War. "Teaching factories" in the form of huge laboratory buildings with large power plants became a hallmark of engineering campuses, as represented by three prominent Purdue buildings: Mechanics Hall, Old Heavilon Hall, and Michael Golden Hall.

William Goss built the first such teaching laboratory at Purdue and opened its doors in 1879. He said:

> During the first three years of Purdue's existence the task of finding out what it should be proved insurmountable, and during this period of uncertainty Purdue was operated under

a plan not materially different from that of other colleges in the state. It was not until 1879 that the trustees made a plunge. They announced for that year a course in Agriculture and a course in Mechanic Arts, and as it happened, I, a youth not yet twenty appeared in the fall of that year to clothe in flesh the conception of a new faith so far as the Mechanic Arts were concerned.[1]

Goss used practical mechanics as a foundation for a comprehensive engineering program that would become one of the world's largest. It would fulfill President Emerson White's plan to make Purdue a shining example of a new kind of industrial college which would take its place beside the classical college in popular esteem. "Its diploma will be honored," he said, "as the evidence of a different but equally fruitful education."[2] The early history of Purdue's engineering program—and the parts played by Goss and White—is recounted in chapter 2.

Another unique and important figure emerged, around 1900, in the person of Frederick Taylor who, like an Old Testament prophet, pointed to the waste and inefficiency in factories and started a crusade for the adoption of what he called "scientific management." Taylor's idea was to have a team of engineers make observations and measurements that could be used to design a factory so that it would run like a well-oiled machine. The Ford assembly line introduced in 1913 became a popular symbol of the efficient factory that Taylor predicted would benefit owners with higher profits, workers with higher pay, and consumers with lower prices.

Management theorist Peter Drucker believed Taylor should get most of the credit for the astounding growth of real income in Europe and America depicted in Figure 1A.

Darwin, Marx, and Freud form the trinity often cited as the *makers of the modern world*. Marx would be taken out and replaced by Taylor if there were any justice in the world. But, that Taylor is not given his due, is a minor matter. It is a serious matter however, that far too few people realize both machines and capital investment were as plentiful before 1880 as they have been since, but there was absolutely no increase in worker productivity during the first hundred years of the Industrial Revolution and consequently very little increase in workers' real incomes or any decrease in their working hours. What made the second hundred years so critically different can only be explained as the result of *applying knowledge to work*.

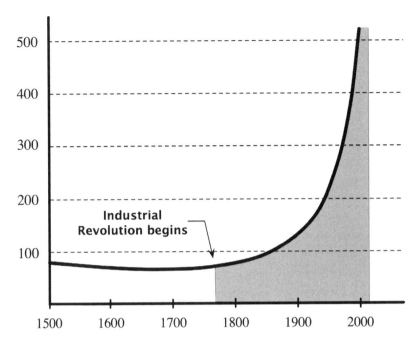

Figure 1A. *Growth of real income per person in England from 1500 to 2000 with a relative income level of 100 in 1860* [3]

Few figures in intellectual history have had a greater impact than Taylor and few have been so willfully misunderstood. Neither the right nor the left will forgive him for proving that capitalism and socialism are irrelevant, only productivity is important, and owners have *duties* not rights, and workers have *responsibilities* not chains. Intellectuals resent his use of *science* for something as trivial as ordinary work because they think work is something slaves do.[4]

Frank and Lillian Gilbreth, two of Taylor's associates, advanced and improved the scientific management movement by addressing the important human questions that it raised. The Gilbreths laid the foundation not only for the academic discipline of industrial engineering but also for the fields of industrial psychology, industrial sociology, industrial physiology, and industrial management. Frank Gilbreth focused on the physiological aspects of work methods and measurement, and Lillian concentrated on the psychological issues of worker participation in the operation and management of industrial enterprises.

The technical and economic benefits of scientific management caused its quick adoption in the industrial buildup before World War I. After the war, the widespread unemployment resulting from the collapse of industry led to European efforts like socialism and communism to revive industry. In America, the engineer Herbert Hoover was elected president leading a campaign to eliminate waste in industry using Taylor's ideas. After the stock market crash of 1929 and his defeat in 1932, Hoover's plan became Franklin Roosevelt's New Deal with more union and government control. Former Secretary of Labor Robert Reich describes the outcome:

> America's prosperity was a product of the alliance between high-volume machinery and large-scale organization which was forged by scientific management. . . . Capital intensive industries like steel, automobiles, chemicals, textiles, rubber, and electrical equipment combined large-scale machine production with scientific management to achieve extraordinary efficiency. . . . America became the envy of the world.[5]

Because of their close friendship with Dean A. A. Potter, the Gilbreths lectured regularly at Purdue for over forty years, guiding the development of the industrial engineering program. As a distinguished faculty member from 1935 to 1947, Lillian worked with Purdue's first professors of industrial engineering, George Shepard and Frank Hockema, to establish an IE program that had a major influence on the teaching of job design and production management. Chapters 3 and 4 examine how industrial engineering developed from the wellspring of scientific management.

In the 1940s, the involvement of engineering schools in the war effort and especially in the postwar national science programs led to a major change in engineering education as it shifted its focus to graduate study and research sponsored by the federal government. As elsewhere, Purdue's undergraduate programs became more science based and oriented toward graduate work. Separate programs in business management and shop technology were developed. Two leaders in that change at Purdue were Frederick Hovde and George Hawkins. Hovde was head of the nation's rocket program at the end of the war, and, when he came to Purdue as president in 1946, he transformed it into a research university. Hawkins followed Potter, becoming Dean of Engineering in 1953.

Hovde and Hawkins began to divide general engineering into three schools—management, technology, and industrial engineering—under separate deans. The plan of study for the BSIE degree introduced in 1955 reflected Hawkins's emphasis on science, systems, and interdisciplinary research and conformed to the new definition of the field that was formulated by the Institute of Industrial Engineers.[6] The transition to the new IE program was made under the first two school heads, Harold Amrine and the author, who led the program for a combined thirty-two years.

Operations research was the most prominent new addition to the curriculum at that time, providing a strong applied mathematics foundation for all of the IE areas and also a fertile area for theoretical research. OR originated in Great Britain during World War II and was first developed in America at MIT, Case Institute, and Johns Hopkins University, led by Phillip Morse, C. West Churchman, and Robert Roy, respectively. Roy used operations research as the foundation for the IE graduate program at Hopkins and his approach was quickly adopted around the country.

Chapters 5 through 8 describe the development of Purdue's IE program in four main application areas: operations research, manufacturing, human factors, and systems engineering. Each chapter begins with an account of national growth in the area and then describes the contributions made by Purdue's faculty in teaching and research. Chapter 9 gives an overview of the professional accomplishments of the School's thousands of alumni, many of whom have made significant technical and social contributions. It is not possible to identify all of them individually, so they are represented by the stories of certain outstanding alumni that are indicative of the character and direction of the practice of industrial engineering.

In chapter 10 we look at the future of the field. By its very nature industrial engineering is future oriented and tries to anticipate how a proposed system may have to adapt to new conditions. This is not an easy task since historians agree that unpredictability is the dominant characteristic of industrial technology, not only in regard to what and when will be the next major breakthrough but how such innovations will ultimately be used by society.

The computer is an outstanding example of technological surprise. In 1952, the author heard John Mauchly tell about building the first functional electronic computer, the ENIAC, at the University of Pennsylvania in 1942 for the Army to calculate artillery ballistic tables. It was the forerunner of the first commercial computer, the UNIVAC, like the one Purdue purchased in 1960. Astonishing as it sounds today, Mauchly said that he and his partner, J. P. Eckert, thought the market for their invention would be very limited because they could think of few other applications that would require such a massive amount of calculations.

After IBM correctly foresaw and won the market for large mainframe computers from UNIVAC, it waited seven years before entering the booming minicomputer market. Later, when IBM dominated the market for desktop computers, they failed to anticipate how important software would become. And, famously, Microsoft did not foresee the popularity of the Web or the search algorithms that Google pioneered.

The American composer John Adams, in his autobiography *Hallelujah Junction*, describes the impact of sound recording on twentieth century musical culture as "so profound that historians have not even begun to analyze its effect."

> The invention of sound reproduction is *the* historical dividing line between the Old World and the New World of music. Recordings, radio transmission, microphones, and loudspeakers radically changed how music is consumed and facilitated the rise of what we now know as popular culture. . . . Even though I heard live concerts from time to time, 90 percent of the virgin experiences I had with the classical canon, not to mention the great works of American jazz, came through the small speakers of a hi-fi set, later replaced by a stereo system. By comparison, an adolescent Aaron Copeland, living at the time of the First World War, could have only have heard a symphonic work by attending a live concert, and chances were slender that the performance would be at the level of a Toscanini recording.[7]

In addition to its impact on listeners, Adams remarks on how new technologies change the creation of music:

> The marriage of the machine to the musical experience is no more and no less than the machine's intrusion into all other parts of our lives. It can, as is often warned, be a source of corruption of the art form. . . . But, at the same time new technologies can be a stimulus for new modes of aesthetic experience and novel creative impulses. Artists should take each new step in the evolution of these machines and turn them into instruments of divine play. It's what we do.[8]

David Nye, in his history of technology development, *Technology Matters*, argues that technology is unpredictable because it is not merely a mechanical process but an expression of a social world: "A fundamentally new invention often has no immediate impact; people need time to find out how they want to use it. Indeed, the best technologies at times fail to win acceptance. Furthermore, the meanings and uses people give to technologies are often unexpected and non-utilitarian."[9] Inventions emerge as the expression of social forces, personal needs, technical limits, markets, and political considerations, as well as from the ideas of a small group of people. Once a basic design is accepted, innovations in production and usage become the dominant determinants of how a technology unfolds.

Nye cites the bicycle as an example of how, after a period of much innovation, the final modern design emerged in 1890, but the cost was equal to the annual salary of an average worker. In the next twenty years, Schwinn and others developed methods of mass production that reduced the price by a factor of twenty, leading to a huge increase in sales with a significant social impact. In 1896, the feminist Susan B. Anthony declared that bicycling had done more to emancipate women than anything else in the world, because the bicycle craze helped kill the bustle and the corset and win social equality. The networks of bicycle roads built in Denmark and the Netherlands on a scale similar to automobile highways built in the United States are an example of how societies create momentum for particular technologies.

The history of industrialization reflects the interaction of social and technical decision making. Henry Ford's success was due as much to his marketing genius as to his technical ability. The fact that European factory workers are paid twice what American workers are paid and given health care is not rooted in technology but in cultural values. When the Internet was created, no one anticipated its use for e-mail. As do other technologies, the Internet creates new businesses, opens new social agendas, raises political questions, and requires a supporting infrastructure. "Cultures select and shape technology, not the other way around," writes Nye. "For millennia, technology has been an essential part of the framework for imagining and moving into the future, but the specific technologies chosen have varied."[10]

The engineer's role in this process is an instrumental one. "Humans wanted experts in technology to achieve society's goals, increase the effectiveness of societal activities, improve the quality of life, enhance human dignity, and develop human capabilities," Gerald Nadler argues.[11] Alan Pritsker writes: "I interpret industrial engineering to be the process of improving total system performance as measured by economic measures, quality attainment, environmental impacts, and how these relate to the benefit of mankind."[12]

In her memoir, *The Quest of the One Best Way*, Lillian Gilbreth observes that her colleagues are "many workers in many fields of activity, far separated. Some of them have nothing apparently at all in common, except the passion for better methods." They are "quest makers," she says, "who long for and search for The One Best Way," as described by her in the following lines taken from *The Quest*. (See Appendix 1B.)

> To most of us, life is to some extent a quest,
> Many seek little things, some a few great things.
> A few seek for *one* thing only; explorers,
> treasure seekers, philosophers, astronomers.
> In the old days, treasure, leisure—the *knight*.
> Today, knowledge, work—the *engineer*.[13]

Gilbreth wrote *The Quest* in 1924, a high point in the period from 1850 to 1950 that Samuel Florman called "the golden age of engineering." In his book, The *Existential Pleasures of Engineering*, Florman, like Gilbreth, extols the personal satisfactions of the engineering profession. But he deplores its decline in popularity that he says began in January, 1950, when President Truman disclosed that work had begun on the hydrogen bomb and Albert Einstein declared that "radioactive poisoning of the atmosphere and hence annihilation of any life on earth has been brought within the range of technical possibilities."[14] Florman writes, "No one who is interested in the engineering profession—least of all those of us who are engineers—could ignore the fall of the engineer from the dizzying heights he once occupied."[15]

What upset Florman most was a growing doctrine that said technology was the root of all evil. Among the antitechnologists, Florman thought Jacques Ellul,[16] Lewis Mumford,[17] Rene Dubois,[18] Charles Reich,[19] and Theodore Roszak[20] were pivotal in forming a movement that claimed:

> (1) Technology is a "thing" or a force that has escaped from human control and is spoiling our lives. (2) Technology forces man to do work that is tedious and degrading. (3) Technology forces man to consume things that he does not really desire. (4) Technology creates an elite class of technocrats, and so disenfranchises the masses. (5) Technology cripples man by cutting him off from the natural world in which he evolved. (6) Technology provides man with technical diversions, which destroys his existential sense of his own being.[21]

"One cannot walk through a mass-production factory and not feel that one is in Hell," W. H. Auden once declared.[22]

Florman acknowledged the good intentions behind the antitechnology movement, but he said it was ultimately a hollow doctrine reflecting a "yearning for simple solutions, where there can be none, and refusing to acknowledge that the true source of our problems is nothing other than the irrepressible human will." He challenged those who would enjoy the benefits of technology but refused to accept responsibility for the consequences by concluding:

> For all our apprehensions, we have no choice but to press ahead. We must do so, first in the name of compassion. By turning our back on technological change, we would be expressing our satisfaction with current world levels of hunger, disease and privation. Further, we must press ahead in the name of human adventure. Without experimentation and change our existence would be a dull business. We simply cannot stop while there are masses to feed and diseases to conquer, seas to explore and heavens to survey.[23]

The fear that technology can be "out of control" was an abiding concern for many thinkers, from Carlyle and Marx to Marcuse and Foucault, who in different ways depict technology as a relentless force reducing people to cogs in the machinery of a totalitarian state. A starting place in this philosophic discussion is the distinction between three basic human activities made by the philosopher Hanna Arendt. She recognized engineering as one form of the creative activity she called *work* or *homo faber* ("man the maker"), which is needed for human survival and growth. It is a purposeful and social process that is intended to answer the basic question, "How?"

Human survival and progress are realized through another basic activity Arendt called *human labor*, which is the personal and instinctive exertion we all do to maintain physical existence. Arendt was critical of Marx for trying to raise human labor to a dominant position in society. Rather, she said, it had to be subordinate to homo faber, which, in turn, is subject to a third basic human activity she called *action*, a political and free activity, that is needed to answer the question, "Why?" Engineering cannot judge the appropriateness of its own technology because it is constrained by commitments to purpose and methodology. For Arendt, the essential quality of political action was absolute freedom of expression.[24]

Arendt's teacher, Martin Heidegger, arguably the most influential philosopher of the twentieth century, thought technology's "relentless, methodical planning" threatened man's ability to understand what it means to be human. But, as the philosopher Andrew Feenberg pointed out, Heidegger allowed no room for technology's evolutionary development. Feenberg notes, "He situates his argument at such a high level of abstraction he literally cannot distinguish between penicillin and atom bombs. Surely, this lack of discrimination indicates problems in his approach."[25] Feenberg adds:

> Technology is not the product of a unique technical rationality but of a combination of technical and social factors. . . . What it is to be an automobile or television is settled by social processes that establish definitions of these objects and grant them specific roles. Technology itself cannot determine the outcome of these processes. . . . Instead of regarding technological progress as a deterministic sequence of developments, we have learned to see it as a contingent process that could lead in many different directions.[26]

I am reminded of my involvement in the operations research studies of large libraries sponsored by the National Science Foundation and described by Philip Morse, head of the Physics Department at MIT in his book, *Library Effectiveness*.[27] Libraries and books are products of the printing press, an invention that had a tremendous, positive influence on society. The same can be said for the extensions of book technology such as the Dewey Decimal and Library of Congress systems of cataloging, shared library depositories, automated checkouts, and computerized catalogs. Potentially even more revolutionary are innovations for online information storage and retrieval such as Web searching and Google Books.

Feenberg saw hope in sociotechnical evolution and the emergence of a necessary public sphere of democratic discussion that is sensitive to technical affairs:

> A good society should enlarge the personal freedom of its members while enabling them to participate effectively in a widening range of public activities. At the highest level public life involves choices about what it means to be human. Today these choices are moderated by technical decisions. What human beings are and will become is decided in the shape of our tools no less than in the action of statesmen and political movements. The design of technology is thus an ontological decision fraught with political consequences.[28]

The eminent French philosopher and University of Chicago professor Jean-Luc Marion concurs with Feenberg, in his recent observation in *Le Monde*:

> The most profound crisis of our era is the one least talked about; the dilution, the evanescence, perhaps even the disappearance of a rationality able to clarify questions that go beyond the mere management and production of objects—questions that decide how we should live and die. Rarely has philosophy or "science" had less to say about the human condition—about what we are, what we can know, what we must do, and what we are allowed to hope for. This dry desert is called nihilism. This is not an opinion that I toss off; it is a fact. And it is our tragedy.[29]

Democratic discussion of technology is consistent with the insights of leading system scientists like Russell Ackoff, who concluded that planning must be experimental and continuous, coordinated internally and integrated externally, but accessible to all stakeholders. In his book, *The Systems Approach*, C. West Churchman discusses the need to deal with many different "anti-planning" viewpoints, such as the skeptics who think planning is meaningless, determinists who think it is hopeless, old pros who prefer to use intuition, fundamentalists who look for guidance in belief systems, pure researchers who abhor planning, and nonintellectuals who simply resist thinking.

To come to a correct understanding, Churchman writes, the systems planner must confront the opposition of the anti-planners. "The systems approach begins when first you see the world through the eyes of another," and learn that there are no experts because every worldview is terribly restricted. "The real expert is still Everyman, stupid, humorous, serious, and comprehensive all at the same time. The public always knows more than any of the 'experts.' . . . The problem of the systems approach is to learn what 'everybody' knows."[30] Or, as Nadler points out, "*Anyone* has a potential to become a valuable contributor in systems planning. The object is to create an atmosphere that fosters the optimal contribution each individual can make."[31]

Japanese auto production is a good example of how technology can adapt and evolve. Unable to justify the cost of Ford-type production methods, Toyota followed the Gilbreths' labor-centric approach and developed a system of "lean manufacturing" that depended on human skill and effort along with mechanization. Toyota fostered a spirit of continuous improvement ("kaizan") by encouraging workers to make millions of suggestions each year.[32] "Most of these ideas are small—making parts on a shelf easier to reach, say—and not all of them work," one critic notes. "But, cumulatively, every day, Toyota knows a little more, and does things a little better, than it did the day before."[33]

Industrial engineers focus on the human interface with technology, which is especially important in the design of mass production and service systems. The presence of people in technical systems is why, as Berkeley professor Paul DeGarmo writes, "industrial engineering occupies an important and needed place, and we have no difficulty in defining, defending, and promoting the uniqueness of our profession."[34] People depend on technology and technology depends on people. The harmonious co-existence of people with technology is an essential part of the quest defined by Lillian Gilbreth.

Industrialization will continue to bring with it new problems and new opportunities. We are only now coming to grips with its impact on the environment and beginning to understand how to make manufacturing sustainable in a world economy. In a 2004 study, *The Engineer of 2020*, The National Academy of Engineering concludes:

> The pace of technological innovation will continue to be rapid (most likely accelerating). The world in which technology will be deployed will be intensely globally interconnected. The population of individuals who are involved with or affected by technology (e.g., designers, manufacturers, distributors, users) will be increasingly diverse and multidisciplinary. Social, cultural, political, and economic forces will continue to shape and affect the success of technological innovation. The presence of technology in our everyday lives will be seamless, transparent, and more significant than ever.[35]

An ultimate solution must include large scale, effective systems for providing society's basic necessities in goods and services through human industry. Nobel Laureate Herbert Simon remarks on the tremendous achievements thus far:

> For over a century in Europe, America, and Japan, output per worker and amount of capital per worker have increased steadily. . . . There has been no long-term trend in percentage of unemployment. Real wages of workers have risen steadily and substantially. . . . The costs that this Revolution sometime imposed on particular groups in society, including labor, have been often acknowledged and often described. Admitting these costs, it is generally agreed that the Industrial

Revolution created for the first time, for mankind, a realistic expectation that abject poverty might be banished from the world.[36]

This is a good way to sum up the progress and purpose of industrial engineers. They are seeking the harmonious co-existence of people with technology—Lillian Gilbreth's enduring quest.

Appendix 1A | The Gilbreth Quest

These excerpts are taken from The Quest of the One Best Way: A Sketch of the Life of Frank Bunker Gilbreth *by Lillian Gilbreth.*

Introduction
To most of us, life is to some extent a Quest, whether we acknowledge it, or not. Many of us seek for numerous easily attained little things— a good time; money enough to buy some small specific thing that we fancy we need or would like—a passing interest or excitement. Some of seek for a few great things—also attainable, but hard to get. A fortune. Fame. The durable satisfaction of life. A few of us seek for one thing only, and that apparently forever unattainable. Those few are those who dedicate their lives to a Quest.

Such are the explorers, who push on and seek new countries and new marvels. Such are the treasure seekers, looking for real or rainbow good. Such are the philosophers, searching out ultimate truths. Such are the astronomers, scanning the heavens for records of the universe. These are the seekers! The Holy Grail, —the Golden Fleece, —the Fountain of Youth, —and now The One Best Way!

In the old days, —treasure divine or earthly. Today, —knowledge. In the old days, —leisure, beneficent or lotus eating; hermit or sybarite. Today, —*Work.* In the old days, —the knight, the cavalier, the romancer. Today, —the *Engineer.* Let us follow one such in his Quest, —A Twentieth Century Adventure! (vii)

The Quest Defined
For the Quest Makers there came a period of discussion, of reviewing the past and its methods, of estimating the factors of the present, and of making plans for the future. All the years up to the time that they had entered into management work, must be carefully considered, and the results must be compared with those years that had passed since that time, both before and during the war.

The comparison made one thing absolutely plain. The decision to enter into the field of management had been right. The work had proved interesting, profitable, and worthwhile. The results were not only satisfactory installations and clients ready with repeat orders, but an accumulating body of data that made each succeeding installation quicker and cheaper for the client, and a much simpler problem for the installer.

Equally right had been the decision that management, while perhaps in its teaching and some of its other aspects is an art, is fundamentally a science and must be conducted as a science, by the laboratory method, and with the most accurate of measurement, with an intensive study of the minutest details. It is true, there has been some criticism that the work placed greater emphasis on motion study than on the installation of a complete plan and mechanism of management, but, as a famous engineer was to say in later years, "It does not matter. They are exactly the same."

The decision having been made, then, that the study and installation of scientific management was their fitting life work, in that it furnished durable satisfaction, they were able at last, after all these many years not only to see but to define the goal of their work. This was to find The One Best Way, fitted to become not only the standard but the super standard. They set as their goal, The One Best Way to Do Work. Derived from all the available information, from all possible study, and as looking forward to an ideal One Best Way, toward which all improvements must be made. (73)

Whither
What will prove success? Not what Frank has accomplished or will be able to accomplish, though this has its place in the final result. Not what this generation accomplishes or may accomplish, though this also has an important place. The proof of the value of the experiment, the real outcome of such a Quest will be its effect upon future generations. (88)

Appendix 1B | Babbage on Manufacturing

Charles Babbage (1792-1871) was the Lucasian Professor of Mathematics at Cambridge University. In his book, On the Economy of Machinery and Manufactures, *he comments on Adam Smith's principle of the division of labor as follows. The reader should note that in the table, Babbage assumes the normal workday is twelve hours for men, women, and children.*

It appears to me that any explanation of the cheapness of manufactured articles, as consequent upon the division of labor, would be incomplete if the following principle were omitted to be stated. That the master manufacturer, by dividing the work to be executed into different processes, each requiring different degrees of skill or of force, can purchase exactly that precise quantity of both which is necessary for each process; whereas, if the whole work were executed by one workman, that person must possess sufficient skill to perform the most difficult, and sufficient strength to execute the most laborious, of the operations into which the art is divided.

As the clear apprehension of this principle, upon which a great part of the economy arising from the division of labor depends, is of considerable importance, it may be desirable to point out its precise and numerical application in some specific manufacture. The art of making needles is, perhaps, that which I should have selected for this illustration, as comprehending a very large number of processes remarkably different in nature; but the less difficult art of pin-making, has some claim to attention, from its having been used by Adam Smith; . . . it will be convenient to present a tabular view of the time occupied by each process, and its cost, as well as the sums which can be earned by the persons who confine themselves solely to each process.

It appears from the analysis we have given of the art of pin-making, that it occupies rather more than seven hours and a half of time, for ten different individuals (4 men, 4 women, 2 children) working in succession on the same material to convert it into a pound of pins; and that the total expense of their labour amounts very nearly to [13 pence], the wages earned by the persons employed vary from 4.5 pence per day up to 6 shillings and consequently the skill which is required for their respective employments may be measured by those sums. . . . If we were to employ for all the processes the man who whitens the pins and who earns 6 shillings per day, even supposing that they could make the pound of pins in an equally short time, yet we must pay him for his time 46.14 pence or about 3s 10d. The pins would therefore cost three times and three quarters as much as they now do by the application of the division of labor. (175-76, 184)

Table 1A. *Babbage's analysis of pin making*

Steps in Pin Making	Done by	Time (hrs/lb)	Cost (pence/lb)	Wage (pence/day)
Drawing	Man	0.3636	1.2500	39
Straighten wire	Woman & girl	0.3000 0.3000	0.2840 0.1420	12 6
Pointing	Man	0.3000	1.7750	63
Twisting & cutting	Boy & man	0.0400 0.0400	0.0147 0.2103	4.5 64.5
Heading	Woman	4.0000	5.0000	15
Tinning or whitening	Man & woman	0.1071 0.1071	0.6666 0.3333	72 36
Papering	Woman	2.1314	3.1973	18
Total		**7.6892**	**12.8732**	

Appendix 1C | The Invention of Mass Production

This is an excerpt from a lecture called "Inventors and Engineers of Old New Haven," in Inventors and Engineers of Old New Haven, *by Richard Shelton Kirby, Yale University, 1939.*

Most intelligent persons know that Eli Whitney invented the cotton gin. Few realize that we owe chiefly to him the modern system of interchangeable manufacture. When the Revolution closed there were no factories in this country for the manufacture of military firearms. . . . In 1798 Whitney undertook to supply for the United States Government 10,000 muskets. He built an armory that was "the most respectable private establishment in the United States for carrying out this important branch of business." It was commenced and carried out "upon a plan unknown in Europe, the object of which was to *substitute the operations of machinery for the skill of the artist.*" In other words, this was the start of interchangeable manufacture, and undertaking requiring foresight and amazing courage.

While the advantage of interchangeability in military arms was obvious and well recognized, no one at that time thought it practical or even possible except at prohibitive cost. Every gun they said would have to be a model and there were not enough skilled workmen to execute the work. Whitney's answer was, instead of building each gun separately, to divide the labor, putting the parts through in lots and transfer the skill of mechanical operations to tools and machinery. In 1801 Jefferson wrote to Madison that Whitney had "invented molds and machines for making all of the pieces of his locks so exactly equal that, take 100 locks to pieces and mingle the pieces, and the 100 locks may be put together by taking pieces which come to hand." Whitney later stated that "it was the understanding that the manufactory should be established on this principle and it has been pursued from the beginning."

During his life Whitney was continually devising new tools and methods which spread elsewhere. The milling machine, one of the most important types of machine tool, can be clearly traced back to him. In the 1913 edition of the *Encyclopedia Britannica* is the statement that the "very first milling machine was running in a Connecticut gun shop in 1818." . . . The armory continued under the management of the Whitney family until it was leased to the Winchester Repeating

Arms Company in 1888. . . . When Colt determined to build his own
armory at Hartford, he planned from the start to embody the principles of interchangeable manufacture in their most advanced form.
(qtd. in Kirby, *Inventors and Engineers*, 6-7)

Appendix 1D | New Lanark

The following description of New Lanark is taken from Robert Dale Owen's 1873 Threading My Way: An Autobiography.

After tens of centuries Arkwright substituted, for human forefinger and thumb, two sets of rollers, revolving with unequal velocity: the lower roller of each pair fluted longitudinally, the upper covered with leather. This gave them a sufficient hold of the cotton as it passed between them. . . . In this way, by an expedient so simple that a child may at a glance comprehend its operation, each set of four rollers took the place of a human being; the metallic fingers, however, working much faster than those of flesh had done. The inanimate spinner with a hundred other similar workmen beside him turned out in a day several times the length of thread the most diligent housewife, toiling at her solitary spinning wheel from morning to night, had been able to produce.

And each company of these *automata* had for its leader or captain, not an adult, male or female, but a child perhaps ten or twelve years old. The urchin learned to direct the ranks of his subordinates with unfailing skill. . . . Thus a tiny superintendent, boy or girl, took the place of a multitude of adult work-people. Myself at the age of twenty-three superintending a manufacturing establishment where some fifteen hundred operative were employed . . . found that each of them, aided by the magical rollers, was producing as much as two hundred cottage spinners had done before Arkwright's day. It need hardly be said that during the first years of such an industrial revolution, the profits in large establishments were very great.

The population of New Lanark in 1784 was upward of seventeen hundred, of who several hundred were orphan children, from seven to twelve years of age; these being procured from the poor-houses of various parishes. It was I believe at that time the largest cotton-spinning establishment in Great Britain, employing about a thousand work-people. The orphan children were comfortably cared for, and but moderately worked; and they attended evening school after the labor of the day was over. . . . Many of the manufacturers of that day urged by the dazzling prospects of fabulous profits, became cruel taskmasters; demanding from the children exertions which even from adults ought never to have been exacted. (11-14)

Appendix 1E | Early Industrial Engineering

In 1503, a Florentine official named Niccolo Machiavelli hired a local engineer named Leonardo da Vinci (left) to study ways of controlling the flow of the Arno River. Leonardo's handwritten notes have survived, like the page reproduced below. It shows some of his calculations for digging a canal. He found that a worker moves a shovelful of earth six arm lengths between the place where he digs to the place where he throws the dirt. One throw takes up to six motions: two or three to load the shovel, one to lift the shovel and turn the body, one to get set and throw, and one to return to the digging place. Some workers, he said, do this in only four motions but they do not last. Working continuously, a good worker can transfer 500 shovelfuls of earth in one hour and a typical shovelful is 10 pounds. He concluded that a good worker could move 5,000 pounds of earth in an hour.

In 1912, Frederick Taylor used similar calculations to demonstrate to members of Congress the basic principles of scientific management. He described how he developed a "science of shoveling" from a motion and time study. He then used this theory to design a production system that cut the work of shoveling in half by employing a special set of shovels.

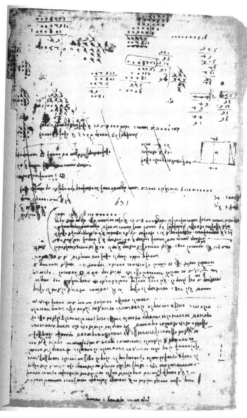

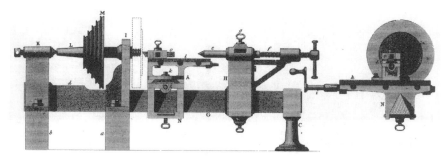

Henry Maudslay launched British precision manufacturing in 1800 with a lathe that combined lead-screw, slide-rest, and change-gears.

Adam Smith (1723-1790)

James Watt (1736-1819) and his double-acting steam engine.

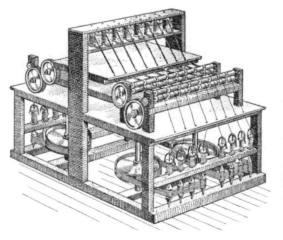

Richard Arkwright (1733-1792) made his cotton spinning machine the centerpiece of the world's first modern factory in 1771.

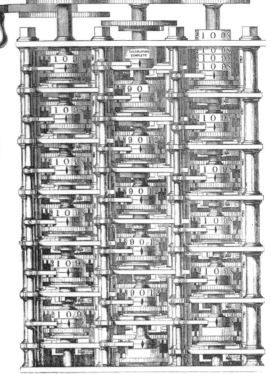

Charles Babbage's (1791-1871) difference engine was the first programmable computer.

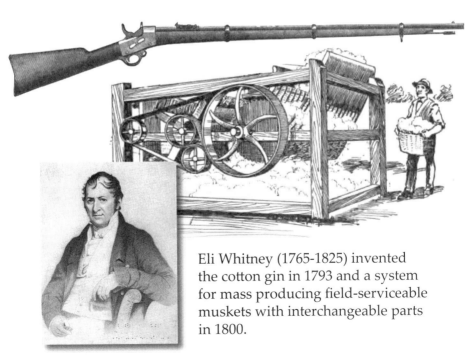

Eli Whitney (1765-1825) invented the cotton gin in 1793 and a system for mass producing field-serviceable muskets with interchangeable parts in 1800.

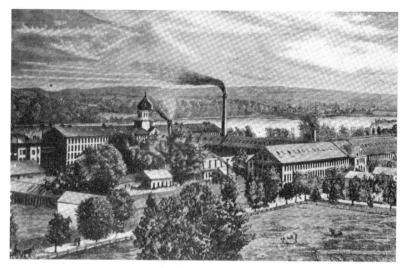

Whitney's idea of mass-production was realized in the Colt Armory of 1857 making 250 arms per day with 600 workmen, 400 tools, and a single steam engine.

In 1825, Robert Owen (1771-1858) showed Congress and Presidents Adams and Monroe his vision of a manufacturing and farming community at New Harmony, Indiana.

Notes

1. Qtd. in Knoll, *Story of Purdue Engineering*, 164.
2. Knoll, *Story of Purdue Engineering*, 16.
3. Clark, *Farewell to Alms*, 47.
4. Drucker, *Post-Capitalist Society*, 39.
5. Reich, *Next American Frontier*, 64.
6. Degarmo, "Industrial Engineering," 70.
7. Adams, *Hallelujah Junction*, 195.
8. Ibid., 209.
9. Nye, *Technology Matters*, 47.
10. Ibid., 210.
11. Nadler, "Role and Scope," 1.3.4.
12. Pritsker, *Papers, Experiences, Perspectives*, 5.
13. Gilbreth, *Quest of the One Best Way*, vii.
14. Florman, *Existential Pleasures*, 12.
15. Ibid., 16.
16. Ellul, *Technological Society*.
17. Mumford, *Myth of the Machine*.
18. Dubos, *So Human an Animal*.
19. Reich, *Greening of America*.
20. Rozak, *Where the Wasteland Ends*.
21. Florman, *Existential Pleasures*, 54.
22. Qtd. in Billington, "In Defense of Engineers," 86.
23. Florman, *Existential Pleasures*, 194.
24. Arendt, *Human Condition*, 177-78.
25. Feenberg, *Heidegger, Habermas*, 2.
26. Feenberg, *Summary Remarks*, 1.
27. Morse, *Library Effectiveness*.
28. Feenberg, *Transforming Technology*, 1.
29. Qtd. in Englund, "How Catholic is France," 12.
30. Churchman, *Systems Approach*, 231.
31. Nadler and Hibino, *Breakthrough Thinking*, 220.
32. Tsutsui, *Manufacturing Ideology*, 138.
33. Surowiecki, "Open Secret," 48.
34. DeGarmo, "Industrial Engineering," 71.
35. National Academy of Engineering, *The Engineer of 2020*, 53.
36. Simon, *Shape of Automation*, 1.

2 | Practical Mechanics

Industrial engineering was first taught at Purdue under the name *practical mechanics* by William Goss, who brought the concept from MIT where he had been a graduate student. Practical mechanics played a key role in fulfilling President Emerson White's vision for Purdue to be a new kind of industrial college and a shining example of the schools mandated by Congress in the Land Grant Act of 1862. White believed that Purdue would fail to fulfill its purpose if its relation to industry was not "close and fruitful." As president of the National Education Association, he had been deeply involved in the long-standing dispute about how to teach engineering in the new land grant colleges. Some thought engineering should be taught in the new science schools being established at the time, but others had serious reservations. The solution to this controversy was a dramatically new approach to higher education found just at the time when Purdue opened its doors for the first time.

The source of practical mechanics can be traced back to 1751 in France with the publication of Denis Diderot's twenty-eight volume *Encyclopedia*, a systematic study of craftsmanship with thousands of carefully drawn illustrations. Peter Drucker wrote:

> It was by no means accidental that articles in the *Encyclopedia* that describe individual crafts, such as spinning or weaving, were not written by craftsmen. They were written by "information specialists": people trained as analysts, as mathematicians, as logicians—both Voltaire and Rousseau were contributors.... The underlying thesis of the *Encyclopedia* was that effective results in the material universe—in tools, processes and products—are produced by systematic analysis, and by the systematic, purposeful application of knowledge. But the *Encyclopedia* also preached that principles that produced results in one craft would produce results in any other. That was anathema to both the traditional man of knowledge and the traditional craftsman.[1]

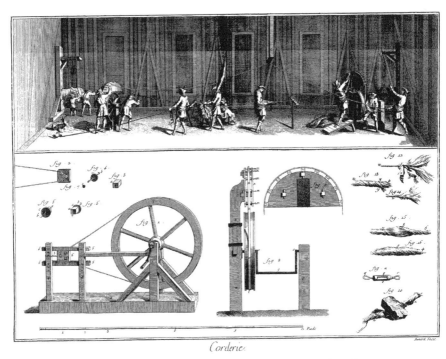

Corderie

Figure 2A. *Illustration from Diderot's* Encyclopedia

Diderot's prodigious work became an instant international success, finding a famous patron in Catherine the Great of Russia. However, because it revealed manufacturing secrets, encouraged industrial innovation, and threatened aristocratic control, the French government suppressed it as seditious.

In 1776, Adam Smith published his study of the dramatic changes taking place in textile manufacturing in England. He observed that factories were highly productive because of the division of labor among several people who, by performing specialized tasks, could greatly multiply the output that any one person could achieve.

> This great increase in the quantity of work, which, in consequence of the division of labour, the same number of people are capable of performing, is owing to three different circumstances; first to the increase of dexterity in every particular workman; secondly, to the saving of the time which is commonly lost in passing from one species of work to another, and, lastly, to the invention of a great number of machines which facilitate and abridge labour, and enable one man to do the work of many.[2]

Smith was describing the early practice of industrial engineering in designing jobs, tools, and work environments so as to expedite the flow of products.

Charles Babbage, the mathematical genius who invented the first programmable computer, wrote a book in 1833, *On the Economy of Machinery and Manufactures*, in which he agreed with Smith that, "the most important principle on which the economy of manufacture depends is the division of labor amongst the persons who perform the work."[3] But, he thought Smith failed to note an important advantage to the owner,

> The master manufacturer, by dividing the work to be executed into different processes, each requiring different degrees of skill and force, can purchase exactly that precise quantity of both which is necessary for each process; whereas if the whole work were executed by one workman, that person must possess sufficient skill to perform the most difficult, and sufficient strength to execute the most laborious, of the operations into which the art is divided.[4]

Babbage theorized that, though the division of labor had been practiced in ancient times, an advanced civilization was needed to realize the advantages of applying the principle in an industrial setting. (See Babbage's calculations on pin making in Appendix 1B.)

In 1879, Oxford economist Arnold Toynbee argued:

> The essence of the Industrial Revolution is the substitution of competition for the medieval regulation, which had previously controlled the production and distribution of wealth. On this account it is not only one of the most important facts of English history but Europe owes to it the growth of two great systems of thought—Economic Science and its antithesis, Socialism.[5]

Toynbee's economic science is the theory of capitalism developed by Smith, Malthus, Ricardo, and Mill to explain how free market capitalism, in theory, should allocate resources in an efficient way. Socialists and communists like Marx and Engels believed the power industrialists wrested from the agricultural aristocrats should go to a worker-controlled and industrial state. Marx ended his *Manifesto* with the call, "Let the ruling classes tremble at a Communistic revolution. The proletarians have nothing to lose but their chains. They have a world to win. Working men of all countries, unite!"[6]

The British historian Raphael Samuel argued that labor power was much more important than capital equipment in making Britain, at mid-century, the "workshop of the world." The mechanization process caused a major change in the pace with which work was done and required a great amount of menial work in the factories and the proliferation of small shops. Hand technology prevailed for skilled tasks. Table 2A, based on Samuel's data on the steam power in 60,000 British factories and shops in 1870, shows that ninety percent of the power was concentrated in twenty percent of the sites that had fifty times more power on average.[7] The abundance of labor caused British industrialists to use labor-intensive methods even when machinery was available. But in the United States, with its sparse immigrant population, mechanization occurred at a much faster rate. Wages were on average three times higher in the U.S. than in Great Britain and labor-saving inventions were necessary for industrial growth. This stark contrast in conditions is reflected in the concerns and accomplishments of two notable pioneers, Robert Owen and Eli Whitney.

Table 2A. *Installed factory steam power in Britain in 1870*

	Factories	Steam Horsepower	Horsepower/ Site
Ironmaking	2,022	221,543	109.57
Textile trades	6,426	414,784	64.54
Bricks, glass, etc	2,887	19,144	6.63
Subtotal	*11,335*	*655,435*	*57.82*
Food manufacture	5,074	7,546	1.43
Metalworking	5,074	51,405	1.01
Woodworking	12,321	8,994	0.73
Clothing, shoes	26,189	2,753	0.11
Subtotal	*48,858*	*70,698*	*1.45*
Total	**60,193**	**726,133**	**12.06**

Owen was a self-made entrepreneur in the booming British textile industry. In 1771, at the age ten, he apprenticed to a textile merchant in London; and eight years later he moved to Manchester where his career took off, first as managing partner in a venture company making cotton spinning machinery; then as manager of a large cotton mill with 500 employees; and finally, at age 29, as sole director of Britain's largest mill complex at New Lanark, Scotland. In Manchester, Owen was the first to import cotton from America. He was a friend of Robert Fulton and a partner in Fulton's steamboat venture. He became deeply involved in Manchester civic and cultural affairs as a member of the Manchester Board of Health and its Philosophical Society.

The New Lanark mills were started by Owen's father-in-law, David Dale, with Richard Arkwright, the inventor of power spinning. (See Appendix 1D.) New Lanark's work force of about two thousand included five hundred orphans from the poorhouses of Glasgow and Edinburgh. Dale was known for his humane treatment of the children but Owen thought their condition was very unsatisfactory and abolished the use of poorhouse labor, employed no child under ten, and provided schools, including an infant school, the first of its kind in Britain. He made large investments in community housing, established a successful company store, and helped found the first British trade union. In 1821, Owen wrote:

> The steam engine and spinning machines, with the endless mechanical inventions to which they have given rise, have inflicted evils on society which now greatly overbalance the benefits which are derived from them. They have created an aggregate of wealth, and placed it in the hands of the few, who, by its aid, continue to absorb the wealth produced by the industry of the many. Thus the mass of the population are become mere slaves to the ignorance and caprice of the monopolists, and are far more truly helpless and wretched than they were before the name of Watt and Arkwright were known.[8]

His radical ideas, and the steady increase in worker productivity and company profits, made Owen famous and New Lanark drew thousands of visitors each year (including heads of state, like the Czar of Russia). He also aroused criticism from all quarters. Employees grumbled about his strict standards of personal behavior; churchmen opposed his ideas of education; politicians thought him too liberal; socialists like Marx thought he was unrealistic; mill owners, including his partners, thought his philanthropies were extravagant.

Owen argued that a well-educated workforce was the key to national prosperity and in the depression following the Napoleonic Wars, he advocated the formation of worker communes as the best way to reorganize society. To prove his point, he sold his interest in New Lanark and used it to finance an experimental worker commune of 3,000 people at New Harmony, Indiana, to which he moved his family. In 1825, after he purchased the land for his new community, he made a presentation to President John Quincy Adams and members of Congress about his plans. As a business venture, New Harmony failed quickly, in part due to the mix of idealists and adventurers who flocked there.

Owen lost a considerable part of his fortune but was undaunted, believing his ideas were far ahead of the times. He returned to England but his five children stayed in New Harmony and became influential Americans. His eldest, Robert Dale Owen, was a congressman, ambassador, writer, and founder of the Smithsonian Institution. His third son, David Dale Owen, was the first official U. S. geologist, and his youngest, Richard Dale Owen, was a noted colonel in the Union Army, a professor at Indiana University, and the first president of Purdue University.

The industrial revolution in textiles and metalwork took root in America in New England, where, for example, the clandestine importation of textile machinery led to the development of towns like Lowell, Massachusetts that emulated New Lanark. Both the textile and metal industries owed much to the pioneering inventions of Eli Whitney. Soon after graduating from Yale in 1792, Whitney visited the South and hit upon the idea of the cotton gin, an ingenious way of processing cotton so that it could be sold in England and New England. Thomas Jefferson told Whitney that his invention was worth more to the United States than the cost of the Louisiana Purchase. Despite this, Whitney went into debt trying to defend his patent rights.

Looking for a new venture, Whitney won a contract to make muskets for the U.S. Treasury at a time when English and French armories had cut off supplies. Although Whitney had never made a gun before and he was unable to find or hire gunsmiths, he invented a revolutionary way of mass-producing guns by dividing the work of a master smith into a series of tasks that could be performed by unskilled workmen using special machine tools invented by Whitney. This was an ultimate realization of Adam Smith's concept of the division of labor. Whitney said, "One of my primary objects is to form the tools so the tools themselves shall fashion the work and give to every part its just proportion."[9] He invented a system for controlling the quality of the parts so that they were interchangeable, readily assembled, and could be repaired in the field. (See Appendix 1C for more on Whitney's system.)

Manufacturing was the centerpiece of world's fairs held in London in 1851, Philadelphia in 1876, and Paris in 1900. The Great London Exhibition of 1851 with its huge cast iron and glass Crystal Palace projected British economic and military superiority. More than 14,000 exhibitors displayed examples of technology developed in the Industrial Revolution. Whitney's company was an American exhibitor, and in a few years the Whitney system was installed at the new Enfield armory in London, the large Colt armory in New Haven, and the huge Beretta armory in Italy that boasted of raw materials coming in one door and guns going out the other. Factories not only changed the way things were made, but they changed the way people learned to work. Traditionally, to do craft work required an apprenticeship in a tightly controlled guild. But in the new industrial factories, rustic young people were taught sufficient manufacturing skills to enable them to satisfactorily replicate and greatly surpass the output of a master craftsman by using special methods and tools.

To deal with the flood of new technology, industrialists and inventors organized local and national societies to exchange information. A society of "civil" engineers was established in England in 1771, to distinguish its members' interests from those of military engineers. From 1830 to 1860 hundreds of voluntary organizations, called mechanics institutes, sprang up in Britain and America in which mechanics of different trades taught each other in defiance of guild exclusiveness. Notable American schools included the Franklin Institute in Philadelphia, the Maryland Institute in Baltimore, and the San Francisco Mechanics Institute. The Ohio Institute was the beginning of the University of Cincinnati. Institutes in Rochester, London, and Manchester also became universities.

The American Society of Mechanical Engineers was formed in 1852; and in the same year, the president of Brown University, Francis Wayland, wrote an essay entitled, *Thoughts on the Present Collegiate System in the United States,* in which he said:

> There has existed for the last twenty years a great demand for civil engineers. Has the demand been supplied by our colleges? The single academy at West Point, graduating annually a smaller number than many of our colleges, has done more towards the construction of railroads than all of our hundred and twenty colleges united.[10]

He resigned his presidency because the Brown faculty would not approve an engineering program.

The demand for engineering education continued to mount, especially in the industrial northern states and, in 1862, Congress passed the Morrill Land-Grant Act, signed by President Lincoln in 1863. The Act called for the establishment of an engineering school in each state of the Union, endowing them with grants of 30,000 acres of public land for each member of Congress from that state. Its implementation was delayed by the Civil War and by local political disputes over where to locate the schools and how to fund them, but the biggest delay was due to the continuing controversy over how the new schools would differ from the existing schools of higher education in each state.

In an 1867 essay, *Our National Schools of Science*, Daniel Colt Gilman, a professor at Yale's School of Science wrote:

> Scientific schools, not classical colleges, are established by the act. The terms of the law, the explanations of its author, the intent of its supporters unite in showing us this beyond doubt. Mathematical, physical, and natural sciences are to be the predominant study rather than language, literature, and history.[11]

However, others argued that there was a fundamental difference between engineering and science that was explained so well recently by Nobel laureate Herbert Simon who said, "Historically and traditionally, it has been the task of the science disciplines to teach about natural things: how they are and how they work. It has been the task of engineering schools to teach about artificial things: how to make artifacts that have desired properties, and how to design."[12] Simon found it ironic that natural science teachers frequently try to drive design from the university curriculum on the grounds that it was not academically respectable or scientifically rigorous.

Important early French contributions to engineering education, other than Diderot's, included the founding of the Ecole Polytechnique as one of the first civil engineering schools, and the introduction of the first teaching laboratory in 1852. Lecturing and demonstrating in laboratories was a radical departure from the one-on-one, tutorial method of classical education. It also differed from the on-the-job instruction in the traditional apprenticeship method of craft instruction.

A crucial breakthrough came in Russia in the 1870s, when lectures and laboratories were combined to create the *Russian method* of technical education, developed by the director of the Russian Imperial Technical Academy in Moscow, Victor Della Vos. His purpose was to train the "mind and hand and eye" to design artifacts by first having students make precise drawings of physical objects that were increasingly more complex, and then having the students produce copies of the objects from the drawings in specially equipped metal and woodworking laboratories. This was done under the direction of a master draughtsman or craftsman who would give lecture-demonstrations.

When Della Vos demonstrated his method at the Philadelphia Centennial Exposition in 1876, John Runkle, the president of MIT, and an accomplished mathematician like Della Vos, was in the audience. He saw that the Della Vos method was an original and transformative answer to the conundrum that had stymied U.S. engineering educators. He immediately changed the curriculum at MIT to follow the Russian method, and, because of what flowed from his bold decision, is now recognized as the leading pioneer of technical education in America. (See Appendix 2C.)

The Philadelphia Exposition happened at a critical time in Purdue's history. After ten years of argument, the Indiana Assembly accepted John Purdue's offer of land and money to establish Indiana's land-grant university in Tippecanoe County. In 1872, the Board of Trustees chose Richard Owen, a geology professor at Indiana University and youngest son of Robert Owen, to be its first president. Owen's plans to have the new school mold the character of its graduates were ridiculed in the press and he resigned after just two years in office. His successor, Abraham Shortridge, served an even shorter time, leaving after only one year for health reasons.

In 1876, the fiery Emerson White, a mathematician like Runkle and Della Vos, became the third Purdue president. He used his inaugural address to describe his dream of building a new kind of "industrial college" at Purdue. He also criticized what other states were doing.

> They are attempting to do the work of the classical colleges, of schools of science, of polytechnic schools and at the same time beat about over a large experimental farm. . . . The relation to industry must be close and fruitful, and the land grant institution that falls short of this fails to do what is most needed for the improvement, not only of agriculture and the mechanical arts, but of all industrial interests and pursuits.[13]

There were to be three academic programs: an academy to prepare students for university-level instruction, a general college of arts and science, and special schools of technology in which the true mission of Purdue would be realized with courses that were practical and idealistic. White continued,

> In all our schemes of education let us not forget that man is more important than his work. The engineer must be swifter than his engine, the ploughman wider and deeper than his furrow, and the merchant longer than his yardstick. In education culture must ever stand before knowledge and character before artisanship.[14]

Purdue differed from the other land-grant schools because of its early emphasis on manufacturing rather than agriculture. White's recruitment of William Goss from MIT to start a school of engineering may have been his shrewdest move. Goss personally built the first shops in Mechanics Hall and announced his commitment to the Russian method in presenting the new school's curriculum in the 1879 catalog.[15]

> SCHOOL OF MECHANICS. The course of instruction and training in this school is based upon the plan designed at the Imperial Technical School of Moscow, Russia, and recently introduced into several American schools of technology. It teaches the student the use of typical hand and machine tools for working in iron and wood, and the elementary principles which underlie all the mechanical trades. It has proved a most efficient substitute for the apprenticeship system, which is fast disappearing. Students taking the course are required to devote two hours each day in work at bench, forge, or machine.

Goss believed practical mechanics was "imperative" to anyone who wanted to be an engineer. Years later, when recalling its beginnings, he said:

> Many people in the State probably assumed that the purpose of the Department of Practical Mechanics was to make good mechanics, and they approved the action because they believed that good mechanics were good men to have in any community. But, the real purpose was to prepare men for a life of usefulness; to have our course in Practical Mechanics to serve in much the same way as those of General Science, or even of Latin and Greek served the old-time college.[16]

In his determination to see that "the new Industrial College will take its place beside the Classical College in popular esteem and its diploma will be honored as evidence of a different but equally fruitful education," President White banned fraternities from the campus because, he said, "progress is choked by customs, practices, and hindrances copied from the classical system, many of these being poor imitations of the stale mummeries of the aristocratic universities of Oxford and Cambridge."[17] In this way, White started a bitter feud with the alumni of Sigma Chi fraternity at Indiana University, who were very influential in State politics. Court battles moved to the floor of the Indiana General Assembly where, in 1883, Purdue's two-year budget appropriation was made contingent on White dropping his antifraternity rule. White resigned in protest but was unable to save the budget and Purdue came close to financial ruin.

President James Smart, who succeeded White, had been superintendent of the Indianapolis school system and, like White, had been a former president of the National Education Association. He was very popular on and off the campus and was called Purdue's "engineering president" because of his enthusiastic support of Goss's program. After adding upper-division degree programs in civil, electrical, and mechanical engineering, Goss became the first dean of engineering in 1890. At the time, there were twenty-eight faculty members: eight in practical mechanics, three in civil engineering (CE), eight in electrical engineering (EE), and nine in mechanical engineering (ME). That year there were 501 students enrolled: 106 in CE, 193 in EE and 202 in ME, all of whom spent their first two years studying practical mechanics.

Old Mechanics Hall, the first engineering building, was replaced in 1894 by the magnificent Heavilon Hall, with drawing rooms for 450 students, a machine shop and foundry in one wing, and a wood shop and forge in the other. Only four days after it was dedicated, however, Heavilon was consumed by fire, and President Smart's response, "I tell you young men that tower shall go up one brick higher," became a Purdue motto.[18] The dedication of the new building, a year later, was Smart's last great achievement before his death in 1900, when he was succeeded by President Winthrop Stone.

In 1890, just after he became dean at Purdue, Goss turned practical mechanics over to Michael Golden, whom he had recruited from MIT. Golden was a legendary figure, both in and out of the classroom, who made an indelible imprint on the University. Knoll said,

He had exceptional ability in fencing, wrestling, tennis, and boxing . . . His devotion to baseball, it was said kept that game from disappearing from the campus. He himself played in the backfield of the faculty football team and in those distant years annually played the seniors. In the last of the faculty-senior games he had the bad luck to suffer a broken leg. It is said that Golden had worked his way through MIT as a sparring partner in Boston boxing parlors and the world champion boxer, John L. Sullivan, said: "He didn't have to be a college professor. If he had stayed with boxing he'd have gone far in the fancy."[19]

Knoll adds that Golden was known as a collector of old books and works of art, a music lover and a skilled performer on the flute, piccolo, and the violin, a photographer, and an inventor whose ingenious machines were used in many manual training schools.

Golden used hundreds of lantern slides that he himself had painstakingly prepared and he delivered his lectures in a rich Irish brogue and with an air of unquestionable authority. He could be arbitrary and threatening: "I'll flunk you if I can and I can if I want to. I'll flunk you just as soon as I can get my pencil out. I don't care if you are right or wrong. If I say you are wrong, you are wrong." Anecdotes about him were cherished as about an epic hero, told and retold and woven into the warmest of campus lore. Freshmen hated him, said the *Exponent*, sophomores respected, juniors appreciated, seniors loved, and alumni reverenced.[20]

Practical mechanics's role at Purdue peaked in 1910 with the erection of the huge Practical Mechanics Building known as "Mike's Castle," and renamed Michael Golden Hall in 1920 when Golden died. His plans for the building, from the *Purdue Engineering Review* of 1910, appear in Appendix 2E. In 1916 James Hoffman, a Purdue alumnus, who had been head of mechanical engineering at Nebraska, became the last head of practical mechanics.

A testimonial to the importance of practical mechanics by a Purdue instructor named William Creighton appeared in 1910.

For the last ten years the number of technical schools has increased ten-fold. We do not have to look far for the cause. In every city, town, and hamlet are springing up machine shops and foundries. Every article around us is machine made. Whence came the designers and constructors of these shops and this machinery? Formerly the designers, proprietors, and superintendents arose from the apprentices, who spent years in the shops hammering iron, in order to learn how to design

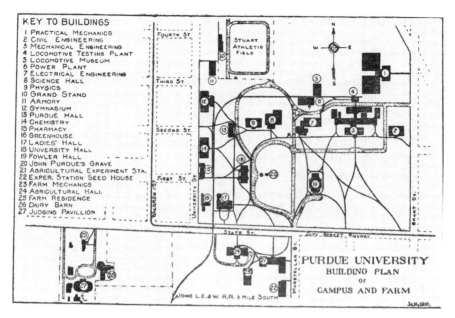

Figure 2B. *The Purdue campus in 1910*

steam engines. For every hour of instruction they received, they worked a week for the proprietor. It took years to learn what systematic training would have given in months. The modern method of education is that pursued by the technical schools. Purdue University has shops filled with full sized tools from the best machine makers in the country. Students are kept in these schools just as long as they are receiving an education therein, and no longer.[21]

Another legendary teacher of practical mechanics was William "Deacon" Turner, who, with Goss and Golden formed the "three mechanics from MIT." He was the long-time graduation marshal until his retirement in 1937 and taught the following courses:

PM 25-26 Shop Work. Practice and lectures in machine tool work in metals. Study of fundamental principles in machine work, such as cutting, speeds and feeds and the shapes of cutting tools: types of machines best suited to a given job. Grinding, gear cutting, screw machine work, and turret lathe work. Elementary production problems in the study of unit or mass production.

PM 134-135 Machine Shop Methods of Production. A continuation of course PM 26 involving more advanced methods in production. Press work in metals, precision work jig, tool, and die making. Tool setting, job analysis, time and cost estimates.[22]

Practical mechanics was the embodiment of Emerson White's ideal American land grant college, a school that produced men who could "do as well as know." Knoll says,

> In retrospect it is easy to see how valuable Practical Mechanics was to Purdue engineering and the whole university. It was the first of the Purdue engineering organizations that lasted, and it had leaders and teachers of exceptional ability, talent, and devotion. For almost sixty years it formed a citadel of engineering strength and placed its mark on all engineering graduates, making them men who could do as well as know, setting standards in industrious habits, and contributing a well-founded spirit to the whole environment.

> It was criticized, often unfairly, for dealing with elementary stuff and did not often receive the credit it deserved, especially since it had to withstand the erosion that came with changing times. But it did its job faithfully and well and men like Mike Golden and his associates made a contribution to education that still deserves to rank with Purdue's greatest.[23]

"Industrial Engineering was the direct descendant of the old Department of Practical Mechanics," he writes. It was through the champions of practical mechanics—White, Goss, Smart, Golden, Turner, and Hoffman—that Purdue industrial engineering got its start.

APPENDIX 2A | The Arrival of *Schenectady*

This is an excerpt from H. B. Knoll's History of Purdue Engineering.

The big event of the Goss-Smart program—the most influential event in all Purdue Engineering—was the arrival in September 1891 of the locomotive *Schenectady*—a steaming, smoking, fire-breathing monster to be studied, tamed and, before long, to be revered as a symbol of power and romance. *Schenectady* brought international recognition to Purdue and kept the fires of enthusiasm hotly burning. Purchasing the locomotive was a bold, almost reckless venture, for there was no guarantee, other than Goss's capacity, that any good would come of it. When it was suggested that the available funds, amounting to $8000, might better be spent for ordinary apparatus, [President Smart] replied that his policy was to secure the big things when he could, because small things would come later without special effort.

Schenectady was delivered to a switch in the tracks of the Lake Erie and Western Railroad, eastward from the site of the airport of the future. A university holiday was declared and a call issued for volunteers to help lift the locomotive from the rails and move it to the campus. . . . In 1891 there were no trucks, tractors, or bull dozers to push or pull. . . . The picture that remains is that of Goss and other staff members, along with a swarm of students, streaming across the fields to the railroad switch, all of them ecstatic with great expectations. In the history of engineering education this was probably the only time when classes were dismissed to celebrate the arrival of a piece of laboratory equipment.

From the distant railroad switch the locomotive was pulled (by three horse teams) across a wheat field and a clover field, over land covered (today) by university housing (on rails mounted on sets of moveable skids). It was brought on to State Street about a block west of the intersection of State and Russell and then driven down State about as far as Grant. It was then turned through something more than a right angle and, avoiding a hollow at the future site of the Memorial Center, was pulled across campus to the newly built Engineering Laboratory.

In the testing plant, *Schenectady* was hitched to a post and its wheels made to run on a kind of treadmill while students and staff studied its performance. *Schenectady* became much more than a locomotive. It symbolized the spirit of a growing university, the aspirations of vigorous young men, and the smoking, steaming, roaring power and might of the rolling trains. It remained at Purdue until the spring of 1898 and, though *Schenectady II* replaced it, its departure was a sad event. Eventually *Schenectady* went into regular service on the Michigan Central Railroad and carried the number 422. It was known among railroad men as "The Schoolmarm." (191-93)

Appendix 2B | Runkle on Mechanic Arts

In his 1882 book, The Manual Element in Education, *John Runkle reported on mechanic arts programs at the University of Maine, MIT, Washington University in St. Louis, and Purdue. He said:*

Before proceeding to a discussion of some schools in which the Russian method of mechanic art education is used, I will simply add that the Imperial Technical School of Moscow was the first to show that it is best to teach art before attempting to apply it; that the mechanic arts can be taught to classes through a graded series of examples by the usual laboratory methods which are used in teaching the sciences. The ideas involved in the system are, first, to entirely separate the art from the trade; second, to teach each art in its own shop; [and] adopt the proper tests for proficiency and progress.

His description of the Purdue program is as follows:

In 1879 the collegiate department of the university was re-organized and made to embrace three courses—the scientific, the agricultural, and the mechanical. An "experimental station" was attached to the agricultural course, and workshops to the mechanical course. Mr. William F. M. Goss, a graduate of the School of Mechanic Arts of the Massachusetts Institute of Technology, was appointed instructor in mechanics. Mr. Goss has furnished me with a plan of the shops and a very full account of the equipment and method of instruction pursued in them from which I have made the following condensed statement.

Applicants for admission to the mechanical course must be over sixteen years of age and pass satisfactory examinations in the common branches, elementary algebra, including quadratic equations, history of the United States, physical geography, and physiology. Graduates of high schools who hold a certificate of the State Board of Education are admitted without examination.

Freshman year.

1st term: woodworking and carpentry, geometry, industrial drawing.

2nd term: pattern making, casting, foundry, geometry, industrial drawing.

3rd term: vise-work, machine drawing, geometry, algebra, industrial drawing.

English composition, one lesson a week; military tactics, three exercises a week.

Sophomore year.

1st term: forging, machine drawing, higher algebra, ancient history.

2nd term: machine-work, machine drawing, trigonometry, physics.

3rd term: machine-work, mill work, machinery, surveying, physics.

Literary exercises, one a week; military tactics, three exercises a week.

The shop practice includes two hours of actual work in the shop daily, five days each week. Students who wish to take courses in mechanical or civil engineering will be admitted to the school on completion of the above courses. The school of engineering will be opened in September 1882.

The shop instruction is divided as follows: bench work in wood, 12 weeks, wood turning, 4 weeks, pattern making, 12 weeks, vise work in iron, 10 weeks, forging in iron and steel, 18 weeks, and machine tool work in iron, 20 weeks.

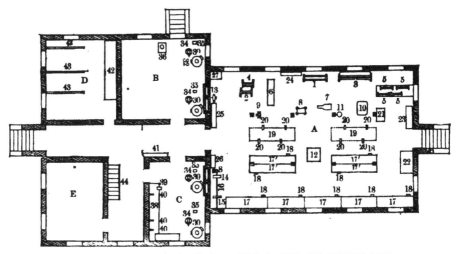

PLAN OF THE MECHANIC ARTS SHOPS, PURDUE UNIVERSITY.

Figure 2C. *Plan of the Mechanic Arts Shop, from John Runkle's book (1882)*

Explanation of the plan. Room A is the main shop for wood and iron work; B and C are forging shops; D is the storage room; E is used by the chemical department. Nos. 1 and 2, machine lathes; 3, wood lathe; 4, speed lathe; 5, four wood turning bench lathes; 6, machine planer; 7, vertical drilling machine; 8, double emery grinder; 9, grindstone; 10, scroll saw; 11, small fret saw; 12, circular saw; 13, Sturtevant blower; 16, shaft running between engine house and shop; 14 and 15, power distributors not shown; 17, wood working benches; 18, vises; 19, iron working benches; 20, vises; 21, instructor table; 22, student work table; 23, 24, 25, 26, tool closets; 27, sink; 28, 29, lathe tool cabinets; 30, forges; 34, anvils; 35, tool racks; 36, forge; 37, forge-work case; 38, bench; 39, vise; 40, iron and steel racks; 41, paint and oil; 42, 43, lumber storage; 44, stairs.

All students are not upon the same work at the same time; but each during his course has an opportunity of learning the use of all the tools and appliances. Nor is a given time allotted to each piece. The slow ones must work extra time to keep up, while those who are quick are given extra work. All students are required to devote all the time allotted to each shop course, and not allowed to pass from one to another in advance of the class, unless they are proficient and can enter a regular class in an advanced branch.

The shop instruction is supplanted by a course of lessons on the theory of the hand and machine tools; more or less use being made of Shelley's *Workshop Appliances*, Rose's *Practical Machinist*, and notes found n the first two volumes on *Building Construction* published by Rivingtons.

I regret that I cannot devote more space to Mr. Goss's interesting account of the good work he is doing at Purdue, which is heartily endorsed by President White. (43-46)

Appendix 2C | Goss on Practical Mechanics

The following statement was written by Dean William Goss in 1920 for an unpublished university history, and is included in H. B. Knoll's Story of Purdue Engineering. *Goss left Purdue in 1907 and went first to be dean of engineering at the University of Illinois and later president of the American Railway Association and the American Society of Mechanical Engineers.*

I think it is extremely important that in any description or discussion of the early work of the Department of Practical Mechanics a serious effort be made to reflect the purpose that was behind it and to trace, as far as may be possible, the evolution of that purpose through the earlier years. It is the spirit of the thing that is worth perpetuating rather than the substance as this may be set forth in lines of practice and courses of study.

Of course the foundation of the purpose is to be found in the terms of the Land Grant Act. Indiana undertook to establish and maintain an institution in the interest of agriculture and the mechanic arts, and the trustees of the university which the state had created were confronted with the problem of determining just what sort of an institution this should be. No definition was supplied as to what a College of Mechanic Arts should teach. During the first three years of Purdue's existence the task of finding out what it should be proved insurmountable, and during this period of uncertainty Purdue was operated under a plan not materially different from that of other colleges of the state. It was not until 1879 that the trustees made a plunge. They announced for that year a course in Agriculture and a course in the Mechanic Arts, and, as it happened, I, a youth, not yet twenty appeared in the fall of that year to clothe in the flesh the conception of a new faith so far as the Mechanic Arts were concerned.

The significance of the action of the trustees is disclosed by a review of the educational thought of the time, trade schools were being talked about. The Imperial Technical Institution of Moscow, Russia, had made a noteworthy exhibit of students' work at the Centennial in Philadelphia in 1876. The Massachusetts Institute of Technology had offered the next year a 'Mechanics Arts Course' based upon the Russian conception. Men of eminence in educational work were giving addresses concerning the educational value of processes underlying the training of the hand. Educators of the old school combatted the innovation, and the debate of the value of hand training as a means to general education was a lively one for a considerable period of years.

The significance of these discussions can best be appreciated by a review of the *Proceedings of the National Education Association*, beginning in the early eighties. I emphasize these discussions because in organizing the work in Practical Mechanics at Purdue there was no expressed purpose in having the University enter upon work in engineering. No engineering course was then offered or predicted. Many people in the State probably assumed that the purpose of the Department of Practical Mechanics was to make good mechanics, and they approved the action because they believed that good mechanics were good men to have in any community.

The Purdue authorities however were all the time proceeding from an educational starting point. The real purpose was to prepare men for a life of usefulness; to have our course in Practical Mechanics to serve in such a process in much the same way that those of General Science or even of Latin and Greek served, except that those of us who stood for the Practical Mechanics entertained a belief that the educational effects of the work which we were projecting were even more significant than those which accrued from the training which characterized the old-time college. We were very likely mistaken in that assumption but the zeal with which we promulgated it was an important factor in our success.

The point I would emphasize is that in the beginning our Practical Mechanics was defended and promoted as a means of general education. Also, that when later the work was matched up with the courses in Engineering, its technical significance assumed other aspects, but the earlier conception was not relinquished. I regard it as one of the real achievements of Purdue that from the beginning the educational quality of the individual courses in Practical Mechanics and in Engineering has been zealously advanced. (qtd. in Knoll, *Story of Purdue Engineering*, 163-64)

Appendix 2D | Practical Mechanics in 1899 *Debris*

The following lampoon of practical mechanics appeared in the 1899 Debris *yearbook.*

The Department of Practical Mechanics is the most practical about the University. It is noted for its practicability. It can be more practical in a minute than any other department in a much longer space of time. Here the student acquires the fundamental principles that will enable him to design a piece of machinery in such a manner that he will not lose job. Also it aids him in sympathizing with the grimy-handed sons of toil, who later in life will look up to him and reverence him as the "boss." There he labors with his own hands to put into practice the ideas previously developed in the drawing room, and thereby gains a knowledge of the details of construction and design that is of great importance in making him a successful engineer.

In the beginning of his course and in his most immature state the prospective master mechanic is found in the foundry (not intended a pun). There he learns to expeditiously and artistically shovel, sift and pound sand and prepare molds from any kind of pattern that grows. This is the mud-pie state of his experience as an engineer. The molds are arranged in rows as in any other shop of importance, and iron at a high temperature is pounded in to fill up the vacant places in the sand. This of course destroys the work of hours spent in making the moulds, but it furnishes employment for the machine shop, and besides gets rid of much old scrap iron.

The wood room is provided with a complete equipment of lathes, sets of carpenter's tools and special machinery for doing the work required. Here the student acquires the knowledge of pattern-making, that is of great advantage to him in designing, and a skill in wood-working that will be of domestic utility in doing odd jobs around the house in later years. Many a man could have lived a truer, nobler life, if with the experience to be gained here he had avoided the profanity of everyday life.

In the machine shop the student learns to turn up the castings made in the foundry with accuracy and a lathe of modern design. He becomes so skilled that he can keep a delicate piece of work going with one hand while he entertains shop visitors with the other.

This department has large and well-disposed equipment, gracefully arranged around the room. It is composed of representatives of the best classes of machine tools and forms a very select and reserved society. One of these tools is a universal grinder, that can grind a piece within a ten-thousandth of an inch with perfect accuracy. But it is sad to think how, with no hope of reward, the patient machine grinds away for hours to reach perfect accuracy and fails to reach its goal when it gets almost close enough to reach out and touch it.

In the forge room, instruction is given in mechanical, not financial forging, as the ignorant may suppose. The faculty having adequate salaries never made a study of the latter method. There is a similarity in the two kinds of work, however, in that mechanical and financial forgers, both usually do time in places where iron bars are a prominent feature of the landscape. The incipient blacksmith learns in the shop to upset a bar of iron, a more difficult process than the initiated may think. He also makes chains, rings and tongs, which he may take home to show his parents how quickly he is acquiring skill and intelligence. The pride of the shop is the home grown trip hammer with which heavy forging is done. We reproduce here camera sketches taken from the department, showing them as we are accustomed to seeing them. The sketches are not life size, but they are accurate, and will in future years remind us of the many pleasant and profitable hours spent in the shops. (56-57)

APPENDIX 2E | Golden on New Laboratories

This article by Michael Golden appeared in the Purdue Engineering Review *of 1910.*

The new laboratory for Practical Mechanics consists of a front in which are located drawing rooms, lecture rooms, and offices, and two wings in which shops are placed. The front is of three stories: the first story having a large lecture room capable of seating three hundred at one time, two recitation rooms and the offices. On the second and third floors are two large rooms.

These are drawing rooms capable of accommodating seventy-five students, so that three hundred students may be accommodated with proper drawing space at one time. The lower drawing rooms are lighted from side windows that are high and take a large portion of the wall space, while the upper drawing rooms, on the top floor, have smaller side windows but are provided with ample top lights. The finish is rough plaster and painted brick. Clothes lockers and closets are on the third story.

The lecture room on the first story is provided with tiered seats and so arranged that a lantern may be used. A hallway runs through the middle of the building on each story, and on the first floor the hallway connects with the shops at the rear. Directly behind the front building is a long hall that is 36 feet wide and will be used for museum purposes. This museum space has its longest dimension running north and south and the basement underneath it is used for lockers, closets, and wash room space. Sets of lockers are provided at each end, one for the freshmen and one for the sophomores.

The shops are to be one story structures with steel roofs, are of what is known as factory construction and are provided with ample top light. They are in the form of two wings, the greater length being east to west. On the extreme north is the wood room, 142 feet by 60 feet. Parallel with this is the foundry 50 by 104 feet. Between these two rooms are located a stock room, storage rooms, and a small lecture room for each shop to be used for demonstration work.

The south wing has corresponding divisions, the larger room being the machine shop and the small room the forge shop, with corresponding minor divisions for the tool rooms, stock rooms, etc. Between these two wings there is a court way that is 40 feet wide by the length of the shops. By this means there is ample sidelight, though the principle lighting is from the top. Under the museum will be located the apparatus for ventilating. The heating and ventilating are from separate systems. The heating will be by direct radiation, with the radiators automatically controlled. The ventilation will be a system by itself and is arranged to be increased or decreased as may be required in the different rooms. By making the heating and ventilating separate problems, problems from each promise to be more satisfactory.

The power will be from electric motors and so arranged that some of the principle machines will have individual drives and the remainder will be driven in small unit groups. In some cases, as with wood room lathes, a unit of 5 H.P. has been taken, while with the larger machines 5 to 10 H.P. motors are used.

The basements under the wood room and the machine room are to be used for the storage of stock. In both the wood room and the machine room the floors are made to be very stiff by cross walls in the basement. Effort is made to have as few belts between the ceiling and the machines as possible as so in the wood room and with a portion of the machine room apparatus the belts will be carried through the floor to the basement, and many of the driving motors will be located in the basement. Lockers for the students' work will be provided around the walls, where these are necessary, and in some of the shops the windows have been placed 6 feet from the floor for this purpose.

The lighting will be in some cases by incandescent lamps. In other places lamps and some Cooper Hewitt lamps will be used. In connection with drawing rooms, provision is made for blueprinting both with sunlight and with machines and there are two photographic rooms where work of that description can be carried on. A special feature will be accessibility to the tool rooms. Provisions are made for demonstrating shop operations with the demonstrating rooms being so arranged that the apparatus from the shop will be carried into the rooms and the seats so tiered that every man in a section of fifty may readily see the work of the demonstrator.

If all the rooms in the building were used to their capacity, seven hundred students could be accommodated at one time. The apparatus from the present shops will be moved during the summer vacation and the new laboratories will be ready for the students when school opens in September. (6-8)

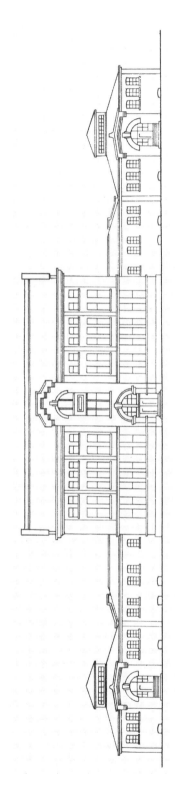

Figure 2D. Part 1. *Plan and Elevation: New Laboratories for Practical Mechanics as drawn by Michael Golden in 1910*

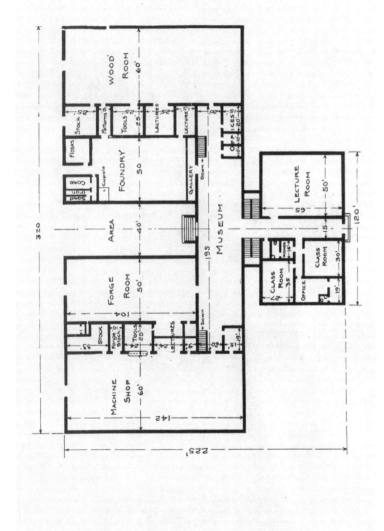

Figure 2D. Part 2. *Plan and Elevation: New Laboratories for Practical Mechanics as drawn by Michael Golden in 1910*

Appendix 2F | Images of Practical Mechanics

Purdue's first presidents: Richard Owen, Emerson White, James Smart.

The Purdue campus in 1900: (left to right) Agriculture Experiment Station, Ladies Hall, University Hall, Mens Dormitory, Electrical Engineering and Heavilon Hall.

A meeting of the Purdue faculty in the library of University Hall around 1898 with William Goss (1), William Turner (2) and President Smart (3).

54 *Chapter Two*

William Goss in 1899, with Mechanics Hall and one of its shops.

Heavilon Hall burning in 1894 and in 1899, rebuilt "one brick higher."

The Schenectady with Goss, far left, and Michael Golden, third from right.

Michael Golden in 1899 (far left); Prof. William Turner in Mechanics Hall, in the 1880s (left).

Engineering students in the forge shop in Heavilon Hall.

Instructor James Hoffman and the wood shop of Heavilon Hall, with a Purdue lathe in the foreground.

The Practical Mechanics Building in 1910 later became Michael Golden Hall in 1920.

From left, Civil (later Grissom Hall), Heavilon Hall (with tower) and Michael Golden Hall (lower right) in 1920.

The manufacturing lab in MGL in the 1930s.

Engineering drawing class in the Practical Mechanics Building in 1910.

Michael Golden with sisters Helen and Katherine, both Purdue graduates and instructors.

Notes

1. Drucker, *Managing for the Future*, 28.
2. Qtd. in Bowditch and Ramsland, *Voices of the Industrial Revolution*, 15.
3. Babbage, *On The Economy*, 169.
4. Ibid., 175.
5. Toynbee, *Industrial Revolution*, 58.
6. Bowditch and Ramsland, *Voices of the Industrial Revolution*, 180.
7. Samuel, "Workshop of the World," 18.
8. Owen, *Report to the County*, 15.
9. Pursell, *Technology in America*, 47.
10. Qtd. in Stratton and Mannix, *Mind and Hand*, 48.
11. Stratton and Mannix, *Mind and Hand*, 255.
12. Simon, *Sciences of the Artificial*, 55.
13. Qtd. in Knoll, *Story of Purdue Engineering*, 10.
14. Qtd. in Knoll, *Story of Purdue Engineering*, 11.
15. Purdue University, *Sixth Annual Register*, 23.
16. Qtd. in Knoll, *Story of Purdue Engineering*, 163.
17. Qtd. in Knoll, *Story of Purdue Engineering*, 13.
18. Qtd. in Knoll, *Story of Purdue Engineering*, 36.
19. Qtd. in Knoll, *Story of Purdue Engineering*, 167.
20. Knoll, *Story of Purdue Engineering*, 168.
21. Qtd. in Knoll, *Story of Purdue Engineering*, 188.
22. Purdue University, *The Fifty-Third Annual Catalog*, 183.
23. Knoll, *Story of Purdue Engineering*, 185.

3 | Scientific Management

Scientific management was what Frederick Taylor called his revolutionary approach to the design and control of industrial operations. Its ergonomic and management implications were greatly extended by two of Taylor's close associates, Lillian and Frank Gilbreth. These three individuals had an impact on manufacturing productivity and the spread of industrialization in the twentieth century that can hardly be overstated. Moreover, they strove to develop a professional engineering approach to the design and management of industrial enterprises. Through their efforts, industrial engineering grew from its origins in practical mechanics, augmented by the systematic viewpoint of scientific management, to become a full-fledged academic discipline.

Frederick Taylor was born in 1856 and grew up in Philadelphia. He was preparing to study law when he decided to become an engineer by beginning as an apprentice machinist in a steel mill. Within eight years he was chief engineer of the mill, having earned an engineering degree through self-study. He also displayed remarkable athletic prowess by winning the doubles competition at the first U.S. Open tennis championship in 1881, and inventing the adjustable tennis net, a spoon-handled tennis racquet to "give the ball a vigorous spin," and a Y-shaped golf putter. He helped introduce overhand pitching to baseball and wrote a book on how to make a putting green that doesn't depend on surface irrigation.

As an engineer at Midvale Steel, Taylor gained national recognition for his design of a large steam hammer and international acclaim for metal-cutting experiments that won him a gold prize at the Paris Exposition of 1900. His most lasting invention, however, was his revolutionary approach to industrial management that was based on a detailed study of every operation in a steel mill. His central thesis was that factory work had to be subjected to the same kind of rigorous engineering analysis and design that was applied to other areas

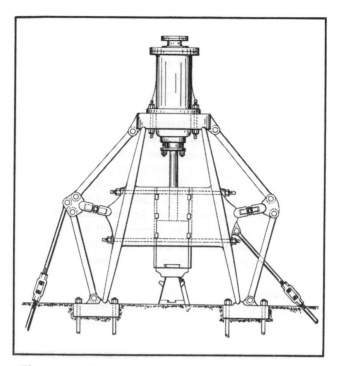

Figure 3A. *Steam hammer designed by Frederick Taylor*

of technology, and that it had to be managed in a professional way for the good of society and not for commercial gain alone.

Taylor first publicized his management ideas in 1903 in a long paper, *Shop Management*. By 1911, when his book, *Principles of Scientific Management* was published, his ideas had become widely known and controversial. In *Principles*, Taylor wrote:

> What the writer wishes particularly to emphasize is that this whole system rests upon an accurate and scientific study of unit times, which is by far the most important element in scientific management.[1]. . . . No one now doubts the economy of the drafting room, and the writer predicts that in a very few years from now no one will doubt the economy and necessity of the study of unit times and of the planning department.[2]

What was needed, Taylor said, was an engineering handbook describing the "correct" methods and times for each operation when done with the best equipment and "first class" workmen.

> The great defect, then, common to all the ordinary systems of management is that their starting point, their very foundation, rests upon ignorance and deceit, and in the one element

which is most vital both to employer and workman, namely, the speed at which work is done, they are allowed to drift instead of being intelligently directed and controlled.[3]

Taylor and his associates became consultants to many companies and government arsenals. He was hired to install his system in the giant Bethlehem Steel Works in Pennsylvania, where he was able to perfect his management ideas and complete the metal cutting experiments that more than doubled steel mill productivity all over the world.[4]

In 1910, Taylor became a national media celebrity when Louis Brandeis, the future Supreme Court justice, solicited Taylor's help in successfully opposing a proposed hike in railroad rates before the Interstate Commerce Commission. After meeting Taylor, Brandeis said, "I quickly recognized that in Mr. Taylor I had met a really great man—great not only in his mental capacity, but in his character—and that his accomplishments were due to this fortunate combination of ability and character."[5] Brandeis won the case with the expert testimony from Taylor's associates, like Harrington Emerson's remark that scientific management could save the railroads $1,000,000 a day.[6] Taylor said the excitement was comparable to that which Teddy Roosevelt inspired in the conservation of our natural resources.

However, scientific management aroused the hostility of labor unions and traditional management, who felt threatened by Taylor's engineering approach. Taylor said:

There is no reason why labor unions should not be so constituted as to be a great help to both employers and men. Unfortunately, as they now exist they are in many, if not most, cases a hindrance to the prosperity of both by forcing their members to work slowly and not allowing a first-class man to do any more work than a slow or inferior workman . . . Men are not born equal, and any attempt to make them so is contrary to nature and will fail. . . . The boycott, the use of force or intimidation, and the oppression of non-union workmen by labor unions are damnable; these acts of tyranny are thoroughly un-American and will not be tolerated by the American people.[7]

A strong response came from James O'Connell, president of the International Association of Machinists, who in a letter to union members in April 1911 said:

The installation of the Taylor System throughout the country means one of two things, either the machinists will destroy the usefulness of this thing through resistance, or it will mean wiping out our trade and organization with the accompanying low wages, life-destroying hard work, long hours and intolerable conditions generally.[8]

Taylor met strong opposition from military line officers for his plan to have professional engineers manage arsenals and naval yards. In the fullness of time, however, all three military academies became engineering schools.

Months after O'Connell's letter, a brief strike by machinists at the Rock Island Arsenal caused the local congressman to have the House of Representatives appoint a special committee to investigate the Taylor system. In his testimony before the committee, Taylor said his system would have only a temporary effect on lowering employment, and he cited Manchester, England, where 5,000 weavers rioted in 1840 because of the power loom, but in 1912 there were 265,000 weavers making 500 yards of cotton cloth for every single yard made in 1840. Part of Taylor's testimony appears in Appendix 3A.

The hearings were sensational and at times came close to fistfights. The journalist Ida Tarbell wrote, "One of the most sportsman-like exhibits the country ever saw was Mr. Taylor's willingness to subject himself to the heckling and the badgering of labor leaders, congressmen, and investigators of all degrees of misunderstanding, suspicion, and ill will."[9] The prominent writer Upton Sinclair asked why Bethlehem workers got only a 60% increase in wages for doing 362% more work and Taylor replied:

Mr. Sinclair sees but one man, the workman; he refuses to see that the great increase in output is also [due] to management. Most of us will see only the workmen and their employers. We overlook the third party, the consumers who ultimately pay both the wages of the workmen and the profits of the employers. . . . In the past one hundred years, the greatest factor tending toward increasing the output and thereby the prosperity of the civilized world has been the introduction of machinery to replace hand labor. And without doubt the greatest gain through this change has come to the consumer. And this result will follow the introduction of scientific management just as surely as it has the introduction of machinery.[10]

One congressman told Taylor privately that when the workmen understood his system, they would rise up and demand it. Nevertheless, Congress passed a law prohibiting the use of Taylor's methods in all federal facilities, including post offices, which lasted thirty-five years. Industrialists, however, were enthusiastic about scientific management and Taylor was the main speaker at efficiency conferences across the country. Sixty-nine thousand people attended one in New York in 1914. In France, Prime Minister Clemenceau ordered every war plant to use Taylor's methods and Michelin made grants to French technical schools to teach scientific management. In Russia, *Pravda* reported Lenin's order: "We should try out every scientific and progressive suggestion of the Taylor system."[11]

Taylor was severely criticized by a generation of union advocates, organizational behaviorists, and industrial psychologists who charged him with insensitivity to basic human needs. He was attacked for treating workers as automatons incapable of managing their own work and needing constant supervision. In time, however, through the efforts of the Gilbreths and others, organized labor dropped its vendetta against scientific management. AFL founder Samuel Gompers addressed the Taylor Society and his successor, William Green, appeared before the Society three times in the 1920s. In 1940, Morris Cooke, one of Taylor's closest associates, and Philip Murray, chairman of the CIO, co-authored a book, *Organized Labor and Production*, that described the kind of union-management cooperation that evolved during the industrial depression of the 1930s from the work of Taylor's successors, such as the Gilbreth team.[12]

Luther Gulick, called the dean of U.S. public administration, recalled hearing Taylor speak to Gulick's class at New York University.

> Frederick Winslow Taylor, to us young researchers, seemed rather cold, somewhat authoritative and aloof, though his emphasis on truly objective analysis and the timing of all processes and motions was most impressive and to us scientific. He orated on his studies even when talking to a few people and waved his stopwatch to make his points, but we will never forget him. The phrase "one best way" however which has been associated with Taylor, was as a matter of fact coined by Frank Gilbreth and not by Taylor. We also met the other half of the Gilbreth team, Lillian.[13]

Frank Gilbreth was born in 1868 and prepped to go to MIT but instead became a bricklayer. Even in later years, he was very proud of his union membership and would make bets on his prowess as a bricklayer. In ten years he became the chief superintendent of a large construction company and then founded his own firm that became internationally known for its innovative building methods. Lillian

Moller was born into the wealthiest family of Oakland, California in 1878, earning a bachelor's degree in 1900 from the University of California, Berkeley, where she was the first woman to ever speak at a U.C. graduation ceremony. She stayed to complete a master's degree in literature and was working on a Ph.D. in psychology when she met Frank Gilbreth in Boston, as she was about to embark on a European tour.

A biographer of the Gilbreths said:

It is extremely difficult to assert with any degree of certainty which was the more important event of the year 1903 for scientific management, the June meeting of the ASME at Saratoga where Frederick Taylor presented his classic paper on *Shop Management,* or the meeting that same month between Lillian Moller and Frank Gilbreth near the Abbey Murals in the Boston City Library.[14]

Lillian later said that the fateful meeting resulted in the "one best marriage."

At first they lived in San Francisco, where Lillian worked on her doctorate and Frank helped rebuild the city after the earthquake of 1906. They jointly wrote a series of books on construction management. In 1910 they relocated to New York for Frank to expand his construction business and Lillian to advance her study of the psychology of management. The Gilbreths joined Taylor's inner circle at the time when Taylor's fame was spreading around the world. Frank described his first impression of Taylor in the following way.

Now, when I first looked into this Taylor plan of scientific management I admit that I fought Dr. Taylor early and often, but Dr. Taylor was very patient and by degrees I found out that what he advocated was applicable to my business. At first it seemed as if there were no parallel between the work in a machine shop, inspecting bicycle balls, and a half dozen other things that I wasn't much interested in, and the work of the contracting engineer; but after Dr. Taylor had told me about these things, I began to look over several trades, one of which was bricklaying, another concrete work, and still another steel erection, and several other trades that I had worked at with my hands in the past. By degrees I found that everything he said was applicable to my business.[15]

See Appendix 3B for Lillian's description of their first meetings with Taylor.

Their involvement with Taylor led the Gilbreths to decide to quit the construction business and start a management consulting firm. They won an initial contract to oversee the installation of the Taylor system in the New England Butt Company (NEB) in Providence, Rhode Island. Their move to Providence was accompanied by two major upheavals, the death of their child, Mary, and the refusal of the University of California to give Lillian a Ph.D. without another year of residence at Berkeley, even though her committee accepted her thesis. She decided to forego the degree and published her thesis as a book, *The Psychology of Management*, the first by anyone on that subject. In Providence, she completed a second doctoral program at Brown University, in which she applied scientific management to public education.

The contract to install the Taylor management system at NEB took over a year to complete. The Butt Co. originally manufactured cast iron butt hinges, but in the Gilbreths' time made braiding equipment for textile manufacturing. The Gilbreths set up two motion study laboratories: one at home and one at NEB. In the home lab, they studied the motions of surgeons, speed typists, and athletes.

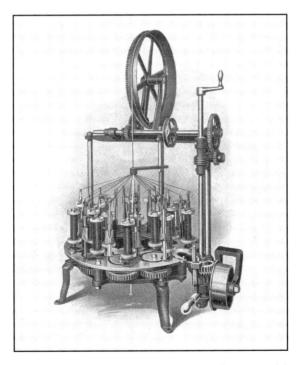

Figure 3B. *Braider made by New England Butt Co., from a 1912 brochure for the Gilbreths' summer school*

Frank even filmed a New York Giants baseball game and observed how a runner on first with an eight-foot lead would have to beat the world track record in order to steal second base. At the Butt Co. they used motion pictures with special timers to study the motions of workers frame-by-frame.

The Gilbreths were able to make substantial improvements at NEB by designing methods and special tools to increase output and quality, and also to reduce fatigue and waste. They introduced a host of "work betterment" ideas like rest periods, worker facilities, career advancement paths, suggestion boxes, a lending library, and meetings to explain their work to union members and the public. Towards the end of the project, "Mr. Taylor was invited to make a thorough inspection, and he showed himself completely satisfied with the results."[16] Also, in Providence, the Gilbreths organized a summer school for college professors that attracted teachers from many engineering schools over a period of three years. This had a major influence on the teaching and the practice of their approach to scientific management and to the development of courses of instruction in industrial engineering.

Frank Gilbreth and NEB president John Aldrich presented their accomplishments at a historic ASME meeting in 1912, where every important person in the scientific management movement was present, except for Lillian, because women were not allowed in the ASME at that time. At the meeting, Taylor openly expressed his opposition to Frank Gilbreth's claim that motion analysis was superior to time study, starting a feud between Taylor and Gilbreth partisans that persisted beyond Taylor's unexpected death in 1915.

Worker participation in shop management was one of the issues that divided the Gilbreths from Taylor. Near the end of his life, Taylor was challenged often for his opposition to collective bargaining by unions over work standards. In one instance, Taylor publicly disputed future Supreme Court Justice Felix Frankfurter's claim that workers should participate in setting shop laws. Taylor held that such rules cannot be found through collective bargaining "any more than the determination of the hour at which the sun will rise tomorrow."[17] Experts, he said, must work out such laws. He never relented in his belief that unions were unnecessary.[18] Frank Gilbreth, on the other hand, boasted about his union membership and contended that motion fatigue was more important than speed. Lillian advocated "bottom-up management," because workers knew more than anyone how best to do a job. Figure 3C illustrates a feature of the Gilbreth approach.

PRIZES FOR SUGGESTIONS

We desire to secure improvement in all departments of our business, and to this end have adopted a plan whereby employees and others may have an incentive to make suggestions with the assurance that all such suggestions will have careful and impartial consideration. Should such suggestions prove of value, the suggester will thereby qualify to compete for a series of prizes to be awarded monthly to employees offering the best suggestions.

Suggestions are invited from all classes of employees. No suggestion need be held back because it appears to be of little importance. The simplest ideas are often valuable.

Suggestions lead to promotion and increased value. They show an interest in our work and organization, and a capacity for greater responsibilities. We invite suggestions upon methods or equipment, methods which will cause more speed, economy or better work, and other matters calculated to advance the interests of the business.

RULES COVERING SUGGESTIONS

All suggestions submitted will be under the supervision of Frank B. Gilbreth, personally.

Write your suggestion and mail it to F. B. G. marked "personal."

Suggestions will be considered promptly. For each suggestion that is accepted, the Company will award the suggester the sum of one dollar, which will be sent to the employee when he is notified that his suggestion has been accepted. We will then be at liberty to adopt the suggestion at any time at our option.

PRIZES

We will award monthly the sum of $20.00 for the most valuable suggestions received during the previous month. This amount will be divided as follows:

FIRST PRIZE	$10.00
SECOND PRIZE	5.00
THIRD PRIZE	3.00
FOURTH PRIZE	2.00
	$20.00

METHOD OF AWARDING PRIZES

On the first Monday of each month, employees who have made suggestions of the greatest value during the preceding month, will be awarded prizes in the order of the value of the suggestion.

As soon as the awards are made, the prizes will be paid in cash, and notices will be posted giving the names of the prize winners, together with a brief description of their suggestions.

Per Order

FRANK B. GILBRETH

Figure 3C. *The Gilbreth suggestion plan at the New England Butt Co. in 1912*

Efforts by Taylor and his associates to discredit the Gilbreths' work caused them to develop what Frank called "our own organization and our own writings so that the workers think we are the good exception" who are labor friendly. They openly criticized stopwatch time study as unscientific, inaccurate and arbitrary, claiming their motion study was free of personal error because it was based on a photographic record. The Gilbreths said, "efficiency is best secured by modifying the equipment, materials, and methods to make the most of the man."[19] They were less interested in increasing output and speed and more interested in reducing fatigue and waste. Rather than "humdrum machine tenders," they said workers should be valued as management allies.[20]

Frank Gilbreth claimed, "It is now possible to capture, record, and transfer not only the skill and experience of the best worker, but also the most desirable elements in the methods of all workers."[21] Eventually, he described any manual task in terms of fourteen basic elementary motions, called "therbligs" (therblig is Gilbreth spelled backwards with *th* reversed). The therblig concept (see Table 3A) made it possible to set time standards without a stopwatch study and has been used in many areas of motion analysis, from sports medicine to robotics and computer animation.

Another Gilbreth invention was the "process chart," representing a sequence of work operations with circles for activities and squares

Table 3A. *Gilbreth's Therbligs or Elemental Task Motions*

Productive Therbligs	Grasp, Transport loaded, Transport unloaded, Release load, Use, Assemble
Unproductive Therbligs	Select, Search, Find, Hold, Inspect, Disassemble, Position, Delay

for decisions as shown in Figure 3D. Frank Gilbreth probably drew this when he served in the Army during World War I overseeing the production of training films. In a 1921 paper the Gilbreths wrote:

> The process chart is a device for visualizing a process as a means of improving it. Every detail of a process is more or less affected by every other detail; therefore, the entire process must be presented in such a form that it can be visualized all at once before any changes are made. Changes made without due consideration of all the decisions and all the motions that precede and follow that subdivision will often be found unsuited to the ultimate plan of the operation.
>
> While the process-chart methods will be helpful in any kind of work and under all forms of management, the best results will come, the authors state, only where there is a mechanism of management that will enforce and make repetitive the conditions of the standards.[22]

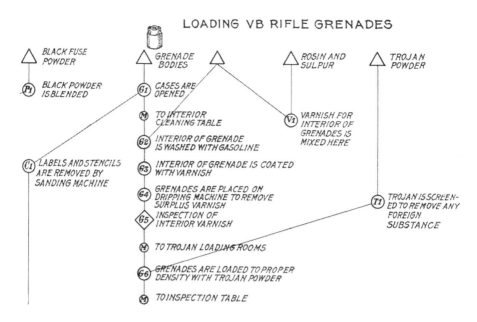

Figure 3D. *Detail from Gilbreth's process chart for loading a rifle grenade*

Still another Gilbreth invention, the "link diagram," was used for graphing travel between work stations. It is believed that Lillian Gilbreth drew the diagram in Figure 3E at Purdue where she worked on the design of institutional kitchens.

After World War I, the Gilbreths built a consulting business in Europe. In 1924, he had just finished a lecture tour in the U.S. (ending at Purdue) and was readying to leave for Prague to give a keynote address at a world congress when he died suddenly of a heart attack. Lillian was left with a large family—eleven children all under eighteen—to support. She took Frank's place on the lecture circuit, organized a management and motion study school out of her home in New Jersey, and developed a successful management consulting business.

An early major client was Macy's in New York, where Lillian gave lectures, redesigned office systems, reorganized the personnel department, and even worked on the sales floor one summer to study retail operations. This led to consulting work for the National Association of Retailers and Sears Roebuck and a lecture tour on retailing at various universities, including Purdue. She made market surveys and designed kitchens and home appliances for General Electric. She wrote a book, *The Home-Maker and Her Job,* arguing that housewives were skilled workers whose work could be managed scientifically.

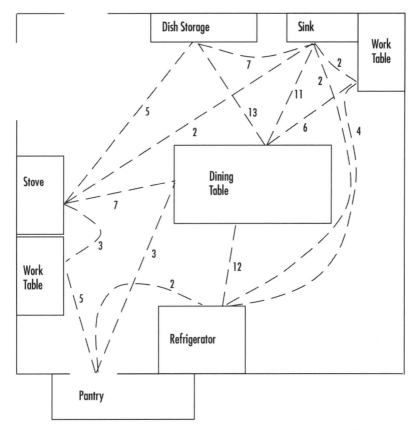

Figure 3E. *Lillian Gilbreth's link diagram for a kitchen* [23]

Homes, she said, should be happy places where individuals achieve fulfillment and freedom by reducing disorder, stress, and fatigue through proper management.

Through her lectures, articles, and radio interviews, Gilbreth became famous for her ability to manage both a professional career and a large family. Following a trip to Europe to give an address on "The Reconciliation of Marriage and Profession," she wrote *Living with Our Children*, part memoir and part child-rearing manual, in which she stressed the importance of teamwork in achieving harmony, adaptability, and a spirit of industry. True cooperation, she said, meant that power had to be delegated and responsibilities shared, despite the loss of efficiency.

American industry, which had boomed during World War I, faltered after the war and collapsed in the Great Depression of 1929. Radical solutions were proposed to solve the social problems of the times, including socialism, communism, and fascism. In America, the economist Thorstein Veblen became a spokesman for many of Taylor's associates when he called for a technocratic revolution.

It is a mechanically organized structure of technical processes designed, installed and conducted by these production engineers. Without them and their constant attention the industrial equipment, the mechanical appliances of industry will foot up to just so much junk. The material welfare of the community is unreservedly bound up with their due working of this industrial system. . . . Right lately these technologists have begun to become uneasily "class-conscious," [with] a growing sense of the waste and confusion in the management of industry [and] a sense of shame and of damage to the common good. So the engineers are beginning to draw together and ask themselves, "What about it?"[24]

Herbert Hoover, the Secretary of Commerce, was the first engineer elected President in 1928 on the basis of his reputation for overseeing relief operations after the war and a famous study of waste in industry. He was a close friend of Lillian Gilbreth and an honorary fellow of the Society of Industrial Engineers. In 1932 he lost his bid for a second term because he failed to obtain a voluntary arrangement between business and labor that would revitalize industry. However, his ideas were revived in the New Deal under Franklin Roosevelt, but with much stronger federal and labor power. In this way, as Robert Reich says, scientific management led to the adoption of policies that favored mass-production industries that could achieve great economies of scale and generate high levels of prosperity.[25]

At the onset of the depression, President Hoover made Gilbreth head of the women's division of the Emergency Committee for Employment, responsible for coordinating the efforts of over 200 organizations. At the Congress of Women at the 1933 Chicago Exposition, where she was a leading speaker, she blamed industry "for putting the machine, not the worker, first." She said, "If wrongly used, the machine can be a peril, but rightly used, it can pull us out of our depression and it can help keep us out of another. A shift from materialism to humanitarianism is imperative."[26]

In 1935, Gilbreth accepted an invitation to teach full time at Purdue, where she and Frank had been regular guest lecturers for many years because of their close friendship with Dean A. A. Potter. At Purdue, she worked with faculty and students in engineering, psychology, and education. She was particularly attentive to women students and became a close friend of Amelia Earhart, who was also on the faculty. Just before going on her fatal flight around the world, Earhart told a friend that, "the most rewarding part of working at Purdue was my connection with Lillian Gilbreth."[27]

Gilbreth kept in close touch with her children who remained in New Jersey with a housekeeper, and she visited them regularly. "Her networking was at a very high level," said her biographer, Jane Lancaster, who found a letter to Eleanor Roosevelt on Purdue notepaper in which she asked if daughter Jane's class could drop by the White House during a trip to Washington. Mrs. Roosevelt invited them all to tea.[28] In the 1950's Gilbreth became famous as the mother in the popular books *Cheaper by the Dozen* and *Belles on Their Toes*, written by two of her children, Frank Jr. and Ernestine. Reportedly she was disappointed because they said very little about her professional accomplishments. Frank Gilbreth was portrayed by both Clifton Webb and Steve Martin as the eccentric father in two very different movie versions of *Cheaper by the Dozen*. An excerpt from *Belles* is recounted in Appendix 3C.

During World War II, Gilbreth worked as a consultant to the War Manpower Commission, WAVES, WACS, and the Brooklyn Navy Yard on the employment of women and disabled persons in war work. She wrote a book on the subject, *Normal Lives for the Disabled*, in 1944. She retired from Purdue in 1948 but continued to make annual visits and give lectures for seventeen more years. In 1964-65 she made the last of many extensive world tours, and gave her final lecture at Purdue. Her health declined steadily until her death in 1972 at age 94.

In 1945, Gilbreth was awarded the prestigious Gantt Gold Medal at a joint meeting of the ASME and the American Management Association. In her lifetime, she received 22 honorary doctorates, including one from Berkeley, 17 honorary society memberships, and 22 gold medals, many from foreign governments, including a prestigious one from Japan, and the Hoover Medal jointly awarded by five engineering societies. She was an advisor to six U.S. presidents. In 1984 the Post Office issued a 40-cent stamp in the *Great Americans* series to honor her.

A 2004 study, *The Engineer of 2020*, by the National Academy of Engineering, ends with a special tribute to Gilbreth for her ingenuity as an engineer.

Lillian Gilbreth is known as the Mother of Ergonomics, a branch of engineering devoted to fitting the workplace to the worker. Ergonomics involves the application of knowledge about human capacities and limitations to the design of workplaces, jobs, tasks, tools, equipment, and the environment. Gilbreth's approach transformed engineering by introducing a primary focus on human needs and capacities.[29]

In her memoir, *The Quest*, Lillian expressed her impatience at resting on laurels when, at the end of the book, she posed a question.

What will prove success? Not what Frank has accomplished or will be able to accomplish, though this has its place in the final result. Not what this generation accomplishes or may accomplish, though this also has an important place. . . . The real outcome of such a Quest will be its effect upon future generations. Whence? Yes, perhaps, as a matter of historical interest. What? Yes . . . But, above all, — whither?[30]

She adds, as the final word, her rallying cry, "The Quest goes on!"[31]

Appendix 3A | The Science of Shoveling

The following is a verbatim excerpt from Frederick Taylor's testimony in 1912 before a special committee of the House of Representatives to investigate "the Taylor and other systems of shop management." The testimony filled three thick volumes, but Taylor's description of the "science of shoveling" is considered a classic piece of American literature.

You may think I am a bit highfalutin when I speak about what may be called the atmosphere of scientific management . . . the intimate and friendly relationship that should exist between employee and employer. I want to emphasize this is one of the most important features of scientific management, and I can hardly do so without going into detail . . . and for this reason I want to take some of your time in explaining the application of the four principles of scientific management to one of the cheaper kinds of work, for instance, to shoveling.

Now, gentlemen, shoveling is a great science compared with pig-iron handling. I dare say most of you gentlemen know that . . . the ordinary pig-iron handler is not the type of man well-suited to shoveling. He is too stupid; there is too much mental strain, too much knack required of a shoveler. . . . You gentlemen may laugh, but that is true; it sounds ridiculous I know, but it is a fact. Now, if the problem were put up to any of you men to develop the science of shoveling . . . where do you think you would begin?

Probably the most important element in the science of shoveling is this: There must be a shovel load at which a first class shoveler will do his best day's work. What is the load? To illustrate: When we went to the Bethlehem Steel Works and observed the shovelers in the yard of that company, we found that each of the good shovelers owned his own shovel; they preferred to buy their own shovels rather than have the company furnish them. . . . We would see a first class shoveler go from shoveling rice coal with a load of 3 1/2 pounds to the shovel to handling ore with 38 pounds to the shovel. Now, is 3 1/2 pounds or 38 pounds the proper shovel load?

What we did was to pick two first-class shovelers. These men were then talked to in about this way: "See here, Pat and Mike, you fellows understand your job all right; both of you fellows are first-class men; you know what we think of you; you are all right now; but we want to pay you fellows double wages. We are going to ask you to

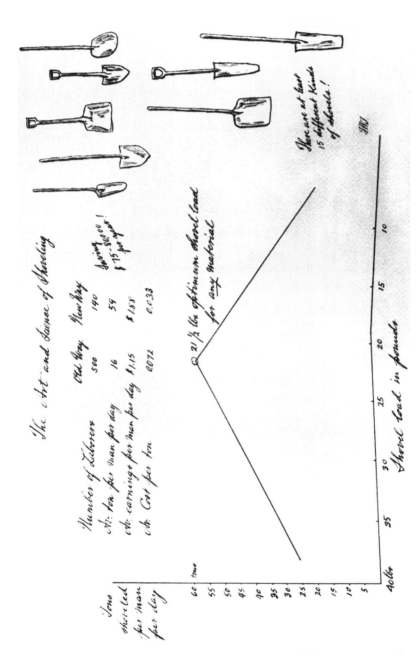

Figure 3F. Taylor's shoveling diagram: *Different jobs require different shovels, but all should hold a load of 21 ½ pounds.*

do a lot of damn fool things, and when you are doing them there is going to be some one out alongside of you all the time, a young chap with a piece of paper and a stop watch and pencil, and all day long he will tell you to do these fool things, and he will be writing down what you are doing and snapping the watch on you and all that sort of business.

The number of shovel loads which each man handled in the course of the day was written down. . . . Our first experiment showed that the average shovel load was 38 pounds and a man handled about 25 tons per day. We then cut the shovel off, making it somewhat shorter, so that instead of 38 it held a load of about 34 pounds. The average of each man went up, and instead of handling 25 he had handled 30 tons per day. . . . It was necessary to furnish each man with a shovel that would hold 21 1/2 pounds . . . This meant keeping 10 or 15 kinds of shovels, and assigning the proper shovel to each one of 400 to 600 laborers in the yard. It meant organizing and planning work at least a day in advance. And, gentlemen, here is an important fact, that the greatest difficulty which we met with in this planning did not come from the workmen. It came from management.

Now, gentlemen, I want you to see clearly that this is one of the characteristic features of scientific management; this is not nigger driving (sic), this is kindness, this is teaching, this is doing what I would like to have done to me if I were a boy trying to learn to do something. The new way is to teach and help your men as you would a brother; to try to teach him the best way and show him the easiest way to do his work. . . . This is the new mental attitude of the management toward the men and this is the reason I have taken so much of your time in describing this cheap work of shoveling. It may seem a matter of very little consequence, but I want you to see, if I can, that this new mental attitude is the very essence of scientific management.

Now all of this cost money. The final test of any system is, does it pay? At the end of three and a half years we had the opportunity of proving whether or not scientific management did pay in its application to yard work. When we got through, 140 men were doing the work of 400 to 600, and these men handled several million tons of material a year, . . . the cost of handling a ton was brought down from between 7 and 8 cents to between 3 and 4 cents. (Taylor, *Scientific Management*, 3:49-67)

APPENDIX 3B | Lillian Gilbreth on Taylor

These are excerpts from The Quest of the One Best Way: A Sketch of the Life of Frank Bunker Gilbreth *written by Lillian Gilbreth as a memorial at the time of Frank's death in 1924.*

Mr. Taylor at this time had already given up the practice of engineering and was doing consulting work and giving his advice to those interested in the movement. He struck one at first glance as a powerful and directing influence, working through other men as well as himself. He was looked up to unquestioningly as leader by those on his staff—Barth, Cooke, Hathaway, and others—who came to feel that, when he spoke, the last word on any possible matter had been said. His was a face that reflected many moods. The eyes, keen and penetrating; the lips thin and tightly closed when he was presenting an argument or meeting opposition—the most genial conversationalist in a sympathetic audience.

Frank plunged into this wholeheartedly. He made himself thoroughly acquainted with the literature. In fact, so thoroughly that he could place almost any reference to *Shop Management* and other classics without looking at the books. . . . [Frank] found that much that he had done was paralleled by what Taylor had done, and that Taylor's method of *timing how long it takes to do work* was new to him, while his method of *studying motions as part of better methods leading to The One Best Way to Do Work* was new to Taylor.

Through their frequent conversations, Mr. Taylor had come to feel that Frank's work readily exemplified scientific management, although it had been done without the direction of Taylor along slightly different lines. . . . [He offered] to cooperate with Frank in writing a book on brick laying. This seemed a most flattering, and in many ways, a desirable offer and was most carefully considered, but Frank and his wife (sic) decided that they preferred to issue their book alone, both because it would then serve its original purpose as an actual system for their organization and because it embodied their viewpoint and their aspirations. This was a most serious decision to come to because it meant or implicitly implied that they had decided to "go it alone" rather than line up as part of the Taylor organization and as followers of his method . . . led to what was really a parting of the ways, in more ways than one.

The decision to enter into the field of management had been right. The work had proved interesting, profitable, and worthwhile. . . . Equally right had been the decision that management, while in its teaching and some of its other aspects an art, is fundamentally a science, and must be conducted as a science, by the laboratory method, and with the most accurate of measurement, with intensive study of the minutest details. . . . They set as their goal, The One Best Way to Do Work, derived from all the available information, from all possible study, and as looking forward to an ideal One Best Way, toward which all improvements must be made. (42-47)

Appendix 3C | Excerpt from *Belles on Their Toes*

Belles on Their Toes *is the sequel to* Cheaper by the Dozen *and was written in 1950 by two of the Gilbreth's children, Frank Jr. and Ernestine. In it they tell the following story about their brother William when he was a student at Purdue.*

On one occasion when Mother was at Purdue, she was asked unexpectedly to speak before a large lecture class in which Bill was enrolled. Bill didn't know about the invitation and picked that particular class to oversleep. Bill's professor told the students that they were fortunate in having a distinguished engineer in their midst. She was Dr. Lillian Gilbreth, and it was gratifying to him that one of Dr. Gilbreth's sons was a member of the class, and doubtless intended to follow in his mother's footsteps.

He cleared his throat and started to call the roll all the way from the A's down to the G's; mother's eyes roved the auditorium, searching for Bill. She was sitting on a platform in a chair next to the professor's table. "Gilbreth?" There was an awkward pause while mother blushed and stopped searching. The professor looked up, cleared his throat again, this time with disapproval, and repeated loudly, "Gilbreth?" A number of Bill's friends sensed the situation simultaneously, and thought they had better come to the rescue. "Here," a dozen voices answered from all corners of the room.

The professor put down his role book and looked bleakly at Mother. He didn't say so, but she gathered the look was intended to convey that he had to put up with a great deal, not the least of which was having Bill as a student. He glared at his audience seeking to find the offenders who had answered to Bill's name. "There seems to be," he said sarcastically, "a good many Gilbreths here today." "The whole family," Mother announced brightly, regaining her poise, and favoring him with her warmest smile. "That's nice." The professor who had seen as much of Bill that semester as he thought he should have didn't think it was nice at all. He licked his pencil and made a show of marking a large zero in his grade book, opposite Bill's name.

Bill spent the remainder of the afternoon and night with Mother, so he didn't see any members of the class during the remainder of the day. Mother didn't mention to him that she had spoken to his group, or that she knew that he had cut class. She thought he was old enough to make his own decisions, and that it wouldn't give him a sense of responsibility if she seemed to be checking up on him. She did spend a good deal of time, though, telling him how she was studying the motions of physically disabled persons, so as to help them find jobs in industry. Bill was interested, and he and Mother looked over her notes and photographs and diagrams of the project.

Bill was a little late, but present, for the lecture class the next morning. He slid into his seat just as the professor finished calling the Cs in the roll book, and was well settled by the time the professor reached the Gs and finally Bill's name. When the professor had run through the list, he told the class he was going to give a written quiz. "I am sure all of you must have learned a great deal from our visitor yesterday," he said. "So today I am going to ask you to write a little summary giving the high points of the talk."

"Who," Bill asked out of the corner of his mouth, "did the old fool drag over here yesterday?" "Are you kidding?" "No, I wasn't here yesterday. I overslept." "You overslept," the boy mimicked. "It was your own mother, you stupid jackass." "Awk," grunted Bill, sinking down in his chair, and wishing he could continue through the floor. Everyone else in the room was writing. You could hear the pens scratching and papers rustling as pages were turned.

Bill hoped that no one would notice that he alone was just sitting there doing nothing. He nudged his neighbor again. "Would you mind telling this stupid jackass," Bill apologized, "what my mother talked about?" "Motion study of the disabled." "Thanks," Bill grinned. He started writing, too. (206-08)

Appendix 3D | Taylor and the Gilbreths

Frederick Taylor

In bowler hats left to right, Henry Gantt, Taylor and Carl Barth inspecting Gilbreth's work at NEB in 1912.

Bethelem Steel Co. executives around 1900. Taylor stands behind third man seated from left.

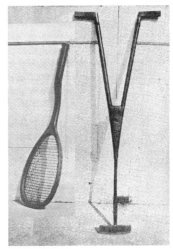

Taylor's "illegal" tennis racket and putter.

President U. S. Grant opening the 1876 Exposition by starting a huge Corliss steam engine.

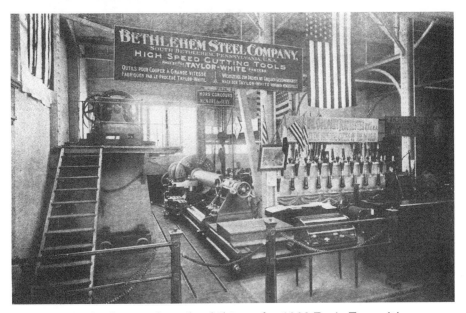

Taylor high-speed steel exhibit at the 1900 Paris Exposition.

Frank and Lillian Gilbreth (right) in 1916 and Lillian (left) at opening of Purdue's Motion and Time Study laboratory with George
Shepard behind projector.

Edward C. Elliott, Lillian Gilbreth and A. A. Potter in 1925.

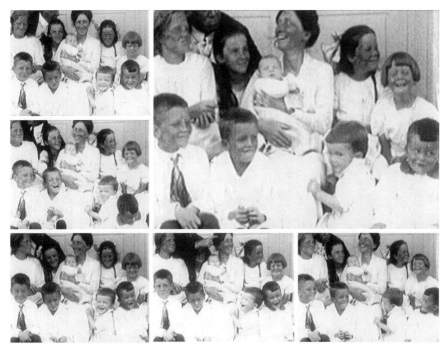

Film of the Gilbreths with nine of their children.

Lillian Gilbreth in 1959.

Lillian Gilbreth demonstrating light ring for filming hand motions.

Left to right, Frank Gilbreth (behind clock), John Aldrich of NEB, and Gantt watch a motion study at NEB in 1912.

Wire models used to study the best way of performing motions.

Gilbreth and Aldrich standing by a braider.

Frank Gilbreth filmed a baseball game between the Giants and the Phillies in 1913.

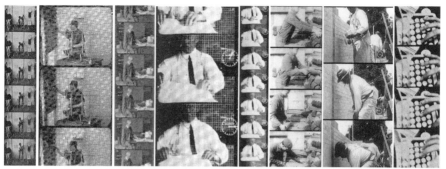

Frames from Gilbreth motion study films: (left to right) man bend-
ing, NEB braiding machine assembler, one-armed secretary, office
worker, pig iron handler, brick layer, typist.

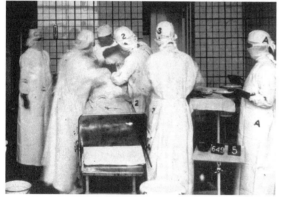

A surgical operation is filmed by Frank
Gilbreth (behind camera, right).

Golfer (right) being photographed with a light ring
on his club.

Lillian Gilbreth at the filming
of a speed typist.

Notes

1. Taylor, *Scientific Management*, 1: 58.
2. Taylor, *Scientific Management*, 1:67.
3. Ibid., 1:45.
4. Copley, *Francis. W. Barkley*, 2:79.
5. Qtd. in Copley, op. cit., 2:370.
6. Copley, op. cit., 7.
7. Qtd. in Copley, op. cit., 2:406.
8. Qtd. in Copley, op. cit., 2:341.
9. Qtd. in Copley, op. cit., 2:341.
10. Qtd. in Copley, op. cit., 2:52.
11. Copley, op. cit., xxii.
12. Cooke and Murray, *Organized Labor*, 137.
13. Van Riper, "Luther Gulick," 6.
14. Lancaster, *Making Time*, 59.
15. Gilbreth, F., *Motion Study*, 195.
16. Gilbreth, L., *Quest of the One Best Way*, 54.
17. Copley, *Frederick W. Taylor*, 420.
18. Copley, *Frederick W. Taylor*.
19. Gilbreth, L., *Psychology of Management*, 16.
20. Gilbreth, L., "Educating the Workers," 1915.
21. Gilbreth, F. and L., *Applied Motion Study*, 141.
22. Gilbreth, F. and L., *Time and Motion Study*, 12.
23. Lehto and Buck, *Introduction to Human Factors*, 107.
24. Veblin, *Engineers and the Price System*, 82.
25. Reich, *Next American Frontier*, 64.
26. Lancaster, *Making Time*, 296.
27. Ibid., 301.
28. Ibid., 302.
29. National Academy of Engineering, *The Engineer of 2020*, 57.
30. Gilbreth L., *Quest of the One Best Way*, 88.
31. Ibid.

4 | Industrial Engineering

The first period in the history of industrial engineering at Purdue started with Goss's practical mechanics program in 1879. In 1920, a second period began, marked by the arrival of two stalwart figures, Edward Elliott and A. A. Potter. In this period Purdue became a hub for teaching the Gilbreth approach to work measurement and production management, attracting faculty and students who would later influence the development of the field. The third and latest period began around 1955, under the leadership of Frederick Hovde and George Hawkins, who advocated a science-based approach to the design of manufacturing and service systems with a strong emphasis on research. In this period, mathematics and computer simulation were applied to a wide variety of technical, commercial, and social problems.

The early need for the field of industrial engineering was voiced in 1886 by the prominent industrialist Henry Towne in a prophetic paper entitled, "The Engineer as Economist," in which he said, "there are many good mechanical engineers; there are also many good business men, but the two are rarely combined in one person."[1] Frederick Taylor's groundbreaking treatise, *Shop Management*, was published in 1903 as a response to Towne's charge. Towne praised *Shop Management* in a graduation address at Purdue in 1905 as the guest of Dean Goss. He said:

> The methods explained and the rules laid down [by Taylor] are probably the most valuable contribution yet made to the literature of industrial engineering. . . . Industrial engineering, of which shop management is an integral and vital part, implies not merely the making of a given product, but making that product at the lowest cost consistent with the maintenance of the intended standard of quality. The attainment of this result is the object which Dr. Taylor has had in view during the many years through which he has pursued his studies and investigations.[2]

See Appendix 4A for more of Towne's remarks.

Early writings about industrial engineering can be found in *The Engineering Magazine*, edited by Charles Going, and in the minutes of the American Society of Mechanical Engineers, of which Towne, Taylor, and Purdue deans Goss and Potter were presidents. The Taylor Society was started in 1910 at Frank Gilbreth's suggestion that "there must be some outfit that will perpetuate Fred Taylor's work. Let us form a society to do it."[3] Going founded the Society of Industrial Engineers in 1915. These two groups merged in 1935 into the Society for the Advancement of Management, forerunner to the American Institute of Industrial Engineers of 1948, which finally became the international Institute of Industrial Engineers in 1981.

At first, Towne and Taylor thought that industrial engineering education could be done with a suitable combination of mechanical engineering and business courses. Before 1900 business courses focused on accounting and personnel issues. Courses in management began to appear in engineering and business programs between 1899 and 1920, as seen in Table 4A. Dartmouth, Harvard, and Wharton created new management schools based on Taylor's ideas. Taylor lectured at Harvard annually, but he repeatedly criticized it and other management programs for their lack of engineering design.

Table 4A. When courses in management were first taught at U.S. universities [4]

Engineering Schools		Business Schools	
MIT	1899	Penn	1901
Penn*	1901	NYU	1903
Cornell	1905	Dartmouth	1904
Iowa	1905	Penn State*	1906
Penn State	1906	Harvard	1908
Purdue	1908	Ohio State	1911
Carnegie	1908	Pitt	1911
Wisconsin	1909	Columbia	1913
Ohio State*	1911	Chicago	1913
Michigan	1914	Michigan	1914
Harvard*	1914	Iowa	1915
NYU	1914	Washington	1917
Dartmouth	1918	N. Carolina	1919
Drexel	1919	Vanderbilt	1919
Pitt	1920		

* Cross-listed course

In 1907 Taylor told the chairman of the board at Penn State that the University would do a great service to industry if it hired Hugo Diemer to teach mechanical engineering from the standpoint of manufacturing. Diemer had written one of the first industrial engineering textbooks, *Factory Organization and Management,* and at Penn State he started a four-year program that he said was superior to what was being offered by business schools like Harvard. Diemer wrote:

When we add to production engineering and production control the planning, forming, and fitting of organizations to work and men to organizations, and write out definite procedures covering the methods and systems involved in this tie-up of production engineering and production control with organizing engineering, we have what is the commonly accepted concept of industrial engineering today.[5]

More examples of Diemer's thoughts are given in Appendices 4B and 4C.

At Purdue Charles Benjamin succeeded Goss as dean and introduced the course *ME 23-24, Industrial Engineering* in 1909 making it a required course for all mechanical engineering seniors. Professor Lawrence Wallace, who was Benjamin's student, took over teaching ME 23-24 in 1910. He attended the Gilbreths' summer school in 1912 and started a popular IE option within ME in 1914. ME 23-24 and a new course, ME 28, Industrial Management, were prerequisites in the option with a capstone course, ME 86-87, Industrial Design, that was taught in part by Benjamin. The catalog descriptions were:

ME 23-24 Industrial Engineering. Lectures on the arrangement and construction of buildings for manufacturing plants; types of buildings, methods of roof construction and lighting and heating plants, arrangement of machinery and installation of cranes and hoists; the organization and management of shops; methods of paying wages; systems of cost accounting and shop bookkeeping; specifications and contracts; economics of power generation and transmission; cost of maintenance and operations; losses due to depreciation, interest, etc.; and total cost of power per horsepower hour per year, including interest, rent, depreciation, labor, fuel, etc.; Lectures illustrated by slides and diagrams.

ME 28 Industrial Management. This course will include lectures and recitations upon the following: (a) Factory lighting. A study of the best sources of lighting for factory purposes,

the quantity if light required, the best location of the lighting equipment, and the best form of shades and fixtures; (b)
Shop sanitation. The various problems concerned with general sanitation of the shop will be studied, such as water and
air purification, etc.; (c) Shop accidents and their prevention.
A general study of the causes and methods of preventing industrial accidents and vocational poisoning. Instruction is
given to the treatment of injuries received in the factory and
special attention is directed to measures for preventing infec-
tion and avoiding permanent disability.

ME 86-87 Industrial Design. The work consists in part of planning a factory layout. . . . A route chart and a model will be
carefully developed, bearing in mind the important factors of
internal transportation, sequence of operations, and general
principles of dispatching. . . . Time and motion studies will be
made of actual operations and these carefully analyzed. Lectures will be given by Dean C. H. Benjamin.[6]

Wallace went to Washington during World War I and on to the University of Texas.

Purdue's IE option was described in an article entitled, "Industrial Engineering at Purdue," by J. E. Hannum that appeared in the
May 1918 issue of the *Purdue Engineering Review*. Hannum wrote:

The Government has asked Industry to supply the many
material requirements of her fighting forces across the seas.
How is Purdue University equipped to meet this present-day
need? A few years ago a course in Industrial Engineering was
begun in a way not at all presupposing, as a senior elective in
the School of Mechanical Engineering. Since that time it has
grown slowly but steadily until at the present time it is a full
senior option.[7]

The second major period in the development of IE at Purdue
began in 1920 with the almost simultaneous arrival of several key
people. Edward C. Elliott succeeded President Stone, A. A. Potter
succeeded Dean Benjamin, and George Shepard replaced Wallace as
the first designated "professor of industrial engineering." It was then
that Potter's close friends, Frank and Lillian Gilbreth, began a series
of annual visits and lectures that continued for over forty-five years.
The Gilbreths, Potter, Shepard, and Shepard's star student, Frank
Hockema collaborated in making Purdue a stage for the development of the Gilbreths' ideas of how industrial engineering could be

developed to achieve the goals of Taylor's scientific management movement.

The Elliott-Potter team oversaw a five-fold increase in every aspect of Purdue's growth over a twenty-five year period. Historian Knoll described how small Purdue must have seemed to Elliott in 1921.

He could look out the window of his office in Eliza Fowler Hall and see almost all there was. In a walk of fifteen minutes he could take a turn about the campus. [However,] had he chosen to inspect the campus in 1945, he would have had about a half day's walk. Under Elliott old Purdue became a new Purdue, larger, stronger, and more diverse. In spirit it became restless and dynamic, never quite satisfied with the progress it made.[8]

Dean Potter's interest in industrial engineering went back to his student days at MIT according to Potter's biographer, R. B. Eckles.

One friendship [Potter] made while he was a student lasted a lifetime. After a lecture by a young bricklayer who had built an MIT building in an unbelievably short time, Potter asked MIT's President Henry S. Pritchett, to introduce him to this genius. Potter was on fire to learn how the "one best way" to lay bricks had been achieved. . . . Thus he met the great motion and time study engineer, Frank B. Gilbreth. Their mutual respect brought Gilbreth, later on, as a lecturer to Kansas State and Purdue. It also led to an affectionate relationship between Gilbreth, his wife Lillian, a great engineer in her own right, and the Potter family. The Gilbreths and the Potters were friends for sixty-three years.[9]

Knoll notes that one of Potter's most memorable professors at MIT was his calculus teacher, John Runkle, the man responsible for the adoption of the Russian method of technical education in America.[10]

Potter concentrated on developing an interdisciplinary program, called "general engineering," that offered courses directly out of the dean's office, in accounting and personnel management. Purdue did not have a business school because of an agreement not to duplicate programs offered at Indiana University. His key advisors were the Gilbreths. Frank was a frequent lecturer at Purdue and, when he died in 1924, Lillian began making regular visits that led to her joining the faculty in 1935 at the urging of both Potter and Elliott. In 1937, a few years after she joined the faculty, practical mechanics was

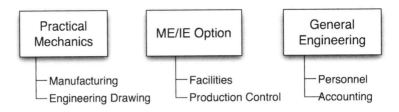

Figure 4A. *IE during the early Elliott-Potter era, 1920-1937*

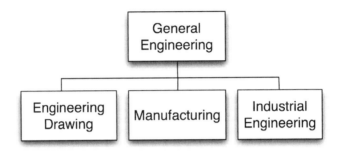

Figure 4B. *IE during the late Elliott-Potter era, 1937-1955*

absorbed into a new Department of General Engineering with three divisions: engineering drawing, manufacturing, and industrial engineering. See Figures 4A and 4B.

George Shepard was recruited in 1919 to take over the courses of Benjamin, who retired. Shepard was an Annapolis graduate with a master's degree from Cornell, who had worked as a consultant for Harrington Emerson's company and was a long-time friend of the Gilbreths. At Purdue, he developed a popular course in industrial management that became the flagship course for Potter's program in general engineering. It was described in the catalog in the following way.[11]

> *GE 110 Industrial Management.* Fundamental principles of management. The application of the scientific method to the selection, control, and disposition of the factors of management—methods, men, materials, etc. Special attention is given to the organization, time and motion study, costs, plant layout, handling materials, personnel administration, production, wage plans, marketing, measurement of management, and the background of industry.

The text for GE 110 was Shepard's book, *The Elements of Industrial Engineering*, in which he tried to reduce industrial engineering to "a few definite and comprehensive principles that students could learn

ALL-INCLUSIVE	PRIMARY	SECONDARY
	Ideals	Adaptation of conditions and work to each other
		$\left\{\begin{array}{l}\text{Correct Methods}\\\text{Instruction}\end{array}\right\}$
Higher Common Sense	Personnel	Fair Deal
		Discipline
		Planning and Despatching
		Records
	Organization	Standards
		Efficiency Reward

Figure 4C. *George Shepard's Principles of Management* [14]

by using them in their daily lives."[12] His principles, shown in Figure 4C, included the basic twelve principles of Harrington Emerson, plus what Shepard called "the one all-inclusive principle of industrial engineering, Higher Common Sense."[13] He said this principle required more than common sense because "sound theory" was needed in both management and technology.

When discussing *Ideals*, Shepard showed his sympathy for Veblen when he said:

> The disastrous labor troubles that have so afflicted this country and others since the World War are striking examples of the evils which result from a conflict of *Ideals* between the management and the mass of industrial privates. The Marxian doctrine that there is by natural economic law an unending conflict between capital and labor, and that the one can prosper only at the expense of the other, is the root of bolshevism. In direct contrast to this, the industrial engineer must offer his doctrine that great wastes are occurring in industry; that by cooperation, labor and capital can prevent these wastes and thereby create new wealth; and by dividing this new wealth equitably between them, both can prosper.[15]

Figure 4D is from the 1920 Purdue *Debris* where Professor Shepard is referred to as the "Admiral." Knoll wrote the following about him.

> Professor Shepard, a retired Navy officer, drew more attention than a four-alarm fire. Calm, factual, and self-possessed, he taught such courses as Motion Study, Time Studies, and Industrial Design and lived what he taught, running his life, it was said, by stopwatch. The registration system that he de-

My instructor is the good Shepard; I shall not pass.

He maketh me to sit up all night; he leadeth me beside the everflowing streams of hot air.

He racketh my soul—he leadeth me into the paths of throwing the bull for his name's sake.

Yea, though I walk through the valley of inefficiency I will fear no labor strikes; for his thirteen principles are with me; his concentration and his adaptations to conditions they comfort me.

He prepareth a schedule card before me in the presence of my classmates; he adorneth my gradecard with hieroglyphics; my note book runneth over.

Surely, Time Studies and Efficiency Rewards shall follow me all the days of my life and I shall dwell in the Insane Asylum forever.

Figure 4D. *A poem and sketch in the 1920 Purdue yearbook* Debris

vised was complained about and successfully used for perhaps twenty-five years. Stories about his personal efficiency were relished all over campus, but if he knew that sometimes people laughed, he gave no sign of concern and indeed was totally indifferent. [16]

There were many stories on campus about Shepard's personal efficiency in "managing his life with a stopwatch" like the following one related by Knoll and involving future dean George Hawkins.

George A. Hawkins, a promising undergraduate, one night walked to town with him. Near the Main Street Bridge, Shepard's briefcase burst open and the papers in it were scattered by a high wind. They lost a precious few minutes retrieving the papers and at Shepard's insistence resumed progress *on a run*. They galloped along Main Street about as far as Sixth before the lost time was regained and Shepard was willing to slow down to a walk.[17]

Shepard's student, Frank Hockema, became the second designated "professor of industrial engineering" at Purdue and was regarded for years as the best teacher on campus. Knoll tells how, because of Hockema's busy schedule, GE 110 had to meet at seven in the morning. He said, "Hockema had no objection to greeting the morning sun, since his custom was to give the university an honest day's work of something like fourteen or fifteen hours. He told his students that getting up early could hardly be considered a hardship in the life of an engineer." Hockema became secretary of the Board of Trustees, executive dean in 1943, and vice president in 1945. Knoll says:

Hockema was a beloved figure among faculty, students, alumni, and people of the community. He was the most accessible man on the campus and one of the friendliest, and when he was felled by a heart attack late in 1955, he received thousands of letters and get-well cards. The meaning of his career is suggested by the fact that he was known as *Mr. Purdue*.[18]

The third prominent teacher of GE 110 was Lillian Gilbreth. In addition to teaching industrial engineering courses, she taught in other departments, notably psychology, education, and home economics. Gilbreth worked with home economics faculty on problems in the design of homes, kitchen, and home appliances and special accommodations for the handicapped that received a great deal of media attention. The third Ph.D. awarded in IE at Purdue was to Orpha Mae Thomas in 1946 for *A Scientific Basis for the Design of Institutional Kitchens*.

Besides Shepard, Hockema, and Gilbreth, the industrial engineering division added professors Harold Amrine, Charles Beese, Robert Field, James Greene, Marvin Mundel, and Barrett Rogers in the 1940s. The division was chaired successively by Beese, Field, Mundel, and Amrine. Amrine gave a detailed account of the history of the IE division and how it evolved into the IE School in his book, *Industrial Engineering at Purdue University: The Roots and First Thirty Years*.

Along with the IE division, general engineering included a manufacturing division, chaired by Roy W. Lindley, that taught courses in foundry, forging, tool and die making, and machining that evolved from practical mechanics. The last members of Dean Goss's "old guard," Professors James Hoffman and William Turner, retired in 1937 along with the name practical mechanics. Profs. O. D. Lascoe, Halsey Owen and Tony Vellinger were colleagues of Lindley and Lascoe replaced him as the leader of the manufacturing faculty.

General engineering also included a drawing division, chaired by Justus Rising, that taught sketching, design, and engineering drawing. Drawing was one of the first courses taught at Purdue in 1874 and was an integral part of the old practical mechanics department. Rising specialized in audio-visual presentation, an area started by Goss and advanced by Golden with his huge lanternslide collection.

Rising took audio-visual instruction to new heights by using movies to teach engineering drawing. Eventually, the drawing faculty moved to Civil Engineering in 1955.

In 1940 Dean Potter chaired the National Engineering Science and Management War Training Program (ESMWT) that oversaw the mobilization of millions of workers in defense industries and about which Peter Drucker wrote:

> Hitler reasoned because the U.S. did not have the large fleet of ships it would not be able to transport enough troops and equipment to Europe. But by applying Taylor's methods, America was able to turn totally unskilled workers into first-rate welders and shipbuilders. "Modern warfare," Hitler further argued, "required precision optics; and there were no skilled optical workers in America." But, with a few months of training, U.S. workers were able to turn out precision optics of better quality than the Germans ever did.[19]

Under Beese's headship, general engineering became a major center for Potter's ESMWT program and its enrollment grew from 3,000 trainees to 60,000. Amrine noted that "Mike's Castle was invaded by hundreds of young women sent by industry for training as engineering aides." After the war, Beese left Purdue and Harold Bolz led general engineering until 1955.

The third period in the evolution of IE at Purdue began at a time when all of engineering was going through a major change. Purdue was growing very rapidly in 1955 under the direction of its new president, Frederick Hovde, who had succeeded Elliott in 1946. He had led the U.S. liaison with Great Britain on science matters and was in charge of the U.S. rocket program at the end of the war. *Time* magazine called him "Purdue's rocket man."[20] Hovde had been a Rhodes scholar, and the first engineer to be Purdue president. (As Minnesota's quarterback, Hovde once ran a fifty yard touchdown against Purdue.) Under Hovde's leadership, Purdue's ranking in output of science Ph.D.s went from nineteenth place to fourth. The Ph.D.s awarded in engineering quadrupled.

George Hawkins, who succeeded Potter as dean in 1953, believed that a "very definite and difficult break" was needed from the way engineering was traditionally taught. He wrote:

> During the 1950's the cumulative pressures of technological change have forced the first-rate engineering schools of the country to make a very definite and difficult break with

the past and to undertake the task of identifying the body of knowledge that is essential for the engineer whose professional life will span the last half of the 20th century. We believe the function of the engineer differs from that of the scientist. In our view engineering encompasses the application of various principles of the sciences as well as background knowledge of the analysis, synthesis, design, and operation of systems and the prediction of their behavior in terms of men, cost, and time.[21]

Hawkins was codirector of a controversial national study, *Goals of Engineering Education,* which recommended eliminating undergraduate degrees, as had been done in medicine and law. Industry successfully opposed the effort, but undergraduate specialization was severely curtailed. At Purdue, Hawkins introduced honors programs to encourage graduate study, established a central statistics consulting service and built the first computing center, containing a room-sized Datatron computer.

This mid-century period brought big changes to industrial engineering. In 1948, a new official definition of industrial engineering was adopted by the American Institute of Industrial Engineers, reflecting the new emphasis on science and systems. It is remarkable because it has remained relevant for over sixty years:

Industrial engineering is concerned with the design, improvement, and installation of integrated systems of people, material, information, equipment, and energy. It draws upon specialized knowledge and skills in the mathematical, physical, and social sciences together with the principles and methods of engineering analysis and design to specify, predict, and evaluate the results to be obtained from such systems.[22]

In 1955, general engineering was renamed the School of Industrial Engineering and Management, with Dean Hawkins as acting head and two departments, industrial engineering and industrial management. The department of industrial engineering was authorized to award the BSIE degree with a plan of study, drawn up by Amrine, that has changed surprisingly little from the current plan that is described in chapter 9.

In 1957, the department of industrial management was combined with the department of economics to form the new School of Management under Emanuel Weiler as dean. IE remained a department for a few years and then became the School of Industrial Engineering in 1961, with Amrine as its first head. A few years later industrial arts, which had survived from the practical mechanics period as a division of the IE program, became a core part of the new School of Technology. George Hawkins remarked on the changes in the 1960 alumni report:

The first students to be graduated by the new Department of Industrial Engineering received degrees this year. The curriculum, a departure from traditional programs, places maximum emphasis on the scientific fundamentals underlying the design of production systems. Much quantitative material formerly taught only to graduate students has been introduced in the undergraduate program, necessarily eliminating some of the art that was formerly taught. A number of students have been attracted to the Ph.D. program established in 1959-60.[23]

Harold Amrine, who was a member of the AIIE definition committee, stepped down as school head in 1969. I succeeded him, and focused the school's effort in four key areas: operations research, manufacturing, human factors, and systems engineering.

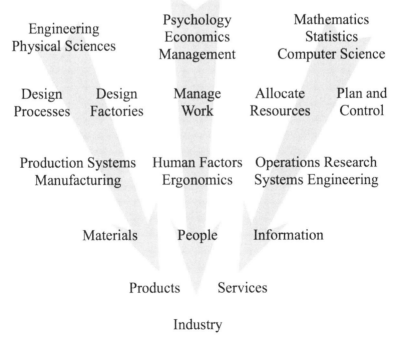

Figure 4E. *Diagram of the scope of the IE program at Purdue*

The BSIE program developed as the starting point for a professional career that could go in many directions. The BSIE courses are kept up to date by being tied directly to the master's level courses. The MSIE program provides expert knowledge in one or more of the specialized areas being developed by thesis research. Projects and grants are sought to support groundbreaking studies by faculty and graduate students.

Theoretical and applied research are done through experimentation, field studies, and software development, often in a cross-disciplinary manner because of the need for expertise in other fields of study. The School attracts faculty with backgrounds in many different areas of engineering and science. Economists and mathematicians are needed to work on optimization methods, psychologists and sociologists on human factors applications, mechanical and materials engineers on manufacturing problems, and electrical engineers and computer scientists on systems hardware and software. This faculty diversity needed in Ph.D. research greatly strengthened the master's degree program and added breadth to the undergraduate program.

Figure 4E diagrams the scope of the program showing how industrial engineering draws on the engineering and physical sciences; on human sciences like psychology and economics; and on mathematics, statistics and computer science. These support the specialized activities of IE in three main areas of manufacturing, human factors, and operations research that come together to coordinate the flows of materials, people, and information in producing products and services. IEs design the production processes and systems. They plan and control the operations and are responsible for how humans are integrated ergonomically into an enterprise as workers and managers.

Since its beginning in 1879, there have been over a hundred faculty members associated with the IE program, as shown in Table 4B. Purdue's first four deans of engineering are listed, including Goss, Benjamin, and Potter, who taught courses in the field, as well as Hawkins, who was acting head of the program in 1955 and had much to do with its growth in the modern era. An alphabetic listing of faculty is given in Appendix 4D.

The professors who were appointed head of the various industrial engineering divisions, departments, and schools through the years at Purdue are listed in Table 4C. The first associate heads were Hewitt Young in 1966 and James Barany in 1969. Barany held the position for over thirty-five years and has advised thousands of IE students, including thousands of beginning freshmen and their parents in the

Table 4B. *Purdue IE faculty by year of appointment*

1879-1919	Benjamin, Golden, Goss, Hoffman, Shepard, Turner, Wallace
1920-1954	Amrine, Balyeat, Beese, Bolz, Field, L. Gilbreth, Greene, Hockema, Hulley, Kirkpatrick, Lascoe, Lindley, Mundel, Pigage, Potter, Radkins, Richardson, Owen, Rogers, Smith, Tilles, Vellinger, Walters
1955-1964	Abhyankar, Adams, Barany, Barash, Bartlet, Brooks, Charnes, Davis, Gambrell, Hartje, Hawkins, Hicks, Leimkuhler, McDowell, Moodie, Randolph, R. Reed, Ritchey, Young
1965-1974	Anderson, Baker, Buck, Deisenroth, ElGomayel, Hill, Mann, Olson, Petersen, Phillips, Pritsker, Ravindran, P. Reed, Reynolds, Roberts, Salvendy, Solberg, Sweet, Talavage
1975-1984	Chandru, Chang, Dolan, Eberts, Fraser, Govinderaj, Johnson, Liu, Meier, Morin, Nof, Rardin, Sadowski, Schmeiser, Sparrow, Stewart, Taaffe, Tanchoco, Wagner
1985-1994	Chandrasekar, Chu, Compton, Coullard, Healy, Koubek, Lehto, Montreuil, O'Cinneide, Papastavrou, Pekny, Prabhu, Shaw, Thomas, Tu, Uzsoy, Wilson, Yih
1995-2008	Caldwell, Cheng, Derjani, Duffy, Engi, Feyen, Landry, Lawley, Lee, Muthuraman, O'Cinneide, Ozsen, Pekny, Richard, Ryan, Stojnic, Stuart, Uhan, Wan, Yi

day-on-campus program. He managed the cooperative education program and started a partnership with AT&T that resulted in the awarding of over 250 master's degrees. Also, for over forty years he trained new faculty advisors for the Institute of Industrial Engineers. In 2005 the associate head duties were divided between Srinivasan Chandrasekar and Jose Tanchoco, and since 2008 Richard Liu has held the position.

In 1989-90, the IE School reached a high-water mark in national and international recognition. At the time, there were six distinguished chaired professors on the faculty: Marshall Professor Shreeram Abhyankar, Ransburg Professors Moshe Barash and James Solberg, Gilbreth Professor Dale Compton, United Technologies Professor William Johnson, and NEC Professor Gavriel Salvendy. Four members of the faculty, Compton, Alan Pritsker, Salvendy, and Solberg, were in the National Academy of Engineering. Johnson was a

Table 4C. *Purdue Industrial Engineering Heads (Including acting and interim division, department and school heads)*

1879–1890	William Goss	1961–1969	Harold Amrine
1890–1910	Michael Golden	1969–1974	Ferd Leimkuhler
1910–1937	James Hoffman	1974–1981	Wilbur Meier
1937–1946	Charles Beese	1981–1993	Ferd Leimkuhler
1946–1955	Harold Bolz	1993–1998	Marlin Thomas
1946–1950	Robert Field	1998–2000	Dale Compton
1950–1952	Marvin Mundel	2000–2005	Dennis Engi
1952–1961	Harold Amrine	2005–2008	Nagabhushana Prabhu
1955–1958	George Hawkins	2008–	Joseph Pekny

fellow of the Royal Academy of Engineering and the Royal Society in England. Five faculty members—Srinivasan Chandrasekar, Tien-Chien Chang, Collette Coullard, Ray Eberts, and Mark Lehto—were recipients of the prestigious Presidential Young Investigator Award conferred by the National Science Foundation.

In 1990 when *U.S. News and World Report* began publishing its annual ranking of graduate engineering programs, Purdue's School of Industrial Engineering was ranked second. In the following two years, 1991 and 1992, it ranked first in the country, and remained near the top for many years. The ranking is based on four factors: faculty research, school budget and resources, student selectivity, and reputation among colleagues across the nation. While the precision of such rankings can be questioned, it indicated that the program was well regarded by its peers. Furthermore, this was the first time that any Purdue school or department in any academic field had achieved a first place national ranking.

The next four chapters describe how the School's programs developed in four areas of concentration. Each chapter begins with a summary of the overall technological development in that area and recounts the contributions made by the School's faculty. In chapter 9, the influence of the School on the practice of industrial engineering is discussed, along with the accomplishments of some of the School's graduates. The overall directions that industrial engineering research has taken and its likely future is described in chapter 10.

Appendix 4A | Henry Towne at Purdue

This excerpt from Henry Towne's talk at Purdue's 1905 graduation ceremony is from his foreword to Frederick Taylor's book Scientific Management.

As pertinent to the subject of industrial engineering, I will quote the following from an address delivered by me, in February 1905, to the graduating students of Purdue University: "The *dollar* is the final term in almost every equation which arises in the practice of engineering in any or all of its branches except qualifiedly as to military and naval engineering, where in some cases cost may be ignored. In other words, the true function of the engineer is, or should be, not only to determine how physical problems may be solved, but also how they may be solved most economically. For example, a railroad may have to be carried over a gorge or arroyo. Obviously it does not need an engineer to point out that this may be done by filling the chasm with earth, but only a bridge engineer is competent to decide whether it is cheaper to do this or to bridge it, and to design the bridge which will safely and most cheaply serve, the cost of which should be compared with that of an earth fill. Therefore, the engineer is, by the nature of his vocation, an economist. His function is not only to design, but also to design so as to ensure the best economical result. He who designs an unsafe structure or an inoperative machine is a bad engineer; he who designs them so that they are safe and operative but needlessly expensive is a poor engineer, and, it may be remarked, usually earns poor pay; he who designs good work, which can be executed at a fair cost, is a sound and usually a successful engineer; he who does the best work at the lowest cost sooner or later stands at the top of his profession, and usually has the reward which this implies."

I avail of these quotations to emphasize the fact that industrial engineering, of which shop management is an integral and vital part, implies not merely the making of a given product, but the making of that product at the *lowest cost* consistent with the maintenance of the intended standard of quality. The attainment of this result is the object which Dr. Taylor has had in view during the many years through which he has pursued his studies and investigations.

About sixty years ago American invention lifted one of the earliest and most universal of the manual arts from the plane on which it had stood from the dawn of civilization to the high level of modern mechanical industry. This was the achievement of the sewing machine. About thirty years ago, American invention again took one of the oldest of the manual arts, that of writing and brought it fairly within the scope of modern mechanical development. This was the achievement of the typewriting-machine. The art of forming and tempering metal tools undoubtedly is coeval with the passing of the stone age, and, therefore, in antiquity is at least as old, if indeed does not outrank, the arts of sewing and writing. Like them it has remained almost unchanged from the beginning until nearly the present time. The work of Mr. Taylor and his associates has lifted it at once from the plane of empiricism and tradition to the high level of modern science.

The conclusions embodied in Dr. Taylor's "Shop Management" constitute in effect the foundations of a new science—"The Science of Industrial Management." . . . Dr. Taylor has demonstrated conclusively that, to accomplish this, it is essential to segregate the *planning* of work from its *execution*; to employ for the former, trained experts and, for the latter, men having the right physical equipment for their respective tasks and being receptive of expert guidance in their performance. Under Dr. Taylor's leadership the combination of these elements has produced in numberless cases, astonishing increments of output and of earnings per employee. (1:6-8)

APPENDIX 4B | Diemer on Industrial Engineering

Hugo Diemer's book, Factory Organization and Management *published in 1910, was one of the first industrial engineering textbooks. These excerpts are from the 1925 edition.*

The industrial engineer must be more than an accountant and more than a production engineer. He considers a manufacturing establishment just as one would an intricate machine. He analyzes each process into its ultimate, simple elements and compares each of these simplest steps or processes with an ideal or perfect condition. He then makes all due allowances for rational and practical conditions, and establishes an attainable commercial standard for every step. The next process is that of developing recorded procedures and organizations to maintain continuously these standards, involving both quality and quantity, and interlocking or assembling of all the prime elements into a well-arranged, well-built, smooth-running machine. It is quite evident that work of this character involves technical knowledge and ability in science and pure engineering that do not enter into the field of the accountant. Yet the industrial engineer must have the accountant's keen perception of money values. The industrial engineer today must be as competent to give good business advice to his corporation as is the skilled corporation attorney. Upon his sound judgment and good advice depend very frequently on the making or losing of large fortunes.

The recognition of industrial engineering as a distinct field is evidenced by the growth of the Society of Industrial Engineers. Still further recognition of industrial engineering as a distinct field is evidenced by the increasing number of colleges and universities establishing departments and distinct courses in industrial engineering. . . . Industrial engineering has suffered a considerable setback in its development not only because of the adoption of the title by under-prepared and inexperienced men, but also by reason of the over-ambitious statement of its scope and aims by some its well-meaning friends and promoters. Some of these have defined industrial engineering as the formulated science of management as applied to any undertaking in which human labor is directed to accomplish any kind of work. . . . This broader field is designated as management engineering although from a standpoint of engineering ethics, the title "engineer" should be used only by a man who attained an engineer's degree.

As soon as Dr. Frederick Taylor's writings and accomplishments were generally known, it became quite apparent that some of the principles developed as an outgrowth of his philosophy of scientific management of industry were capable of universal application. The management of any undertaking, whether it be educational, political, military, religious, financial, mercantile or industrial, can best be developed by following the same principles which Taylor applied to factory management. These principles include analysis, research, measurement of existing organizations, methods and results, standardization, measurement under standardized and improved conditions, control through planning, preparation, scheduling, dispatching and inspection, functional organization, selection and training of workers and equitable payment based on individual efficiency. (6-7)

Appendix 4C | Early Industrial Engineering Schools

These excerpts are from the book A Mental Revolution: Scientific Management Since Taylor *by historian Daniel Nelson.*

The experiences of the Pennsylvania State engineering college made the institution a prototype for other engineering schools. In 1907 Hugo Diemer, an Ohio State graduate who had taught at the University of Kansas became head of the Penn State mechanical engineering department. While a faculty member at Kansas, Diemer had become a devotee of scientific management and a close acquaintance of Taylor's; his appointment [was] due to Taylor's influence. His goal was to make management studies an integral feature of the mechanical engineering program. Penn State already had a course, "Shop Economics," which covered many of the specific features of scientific management. In 1907 Diemer added "Factory Planning," and in 1908 introduced a concentration in "Industrial Engineering" within the mechanical engineering curriculum.

In 1909 he won approval for an industrial engineering department, the first in any American university. Students took conventional engineering courses for their first two years. They studied "Shop Time Study" and "Manufacturing Accounts" as juniors, "Shop Economics," "Labor Problems," and "Factory Planning" as seniors. In 1913 Diemer added "Industrial Management," devoted to "departments and departmental reports, planning, scheduling, time study, labor and efficiency, wage systems, and welfare methods." The following year he added an advanced course on "Scientific Management," which used Taylor's writings as texts. Diemer left Penn State during World War I . . . By 1921 the industrial engineering department had 12 faculty members, six of professorial rank. ...

The experiences of Cornell and Purdue, two other leaders in industrial engineering, were similar. At Cornell, Dexter S. Kimball, . . . introduced a required junior-level courses . . . covering manufacturing methods, cost accounting, and plant management [and] welfare work, wage systems, and labor legislation. In 1914 when Kimball was appointed to head a new department of industrial engineering, [he] introduced a senior course "Industrial Administration," that covered "modern time-keeping and cost-finding systems, methods of planning work and of insuring production, administrative reports, time and motion study, purchasing, etc."

Purdue introduced "Industrial Engineering" in 1908 as a senior requirement for mechanical engineering students. The two-semester course covered factory construction and power generations as well as "organization and management of shops; methods of paying wages; systems of cost accounting and shop bookkeeping. . . ." Charles Henry Benjamin, the dean of the engineering school taught the course until 1910, when L. W. Wallace joined the faculty.

Other universities followed the examples of Penn State, Cornell, and Purdue and created majors for students who had a strong interest in management or a weak interest in "pure" engineering. Several took an additional step and broke the implicit link between engineering and industrial management. . . . By the turn of the century Davis R. Dewey, MIT's distinguished economist, and several colleagues offered course in economics, law, and history, which Dewey wanted to expand into a social science major. He was unable to win the support of the university administration, in part because of an anticipated merger with Harvard. When the merger failed, Dewey succeeded in introducing "Course XV, which combined existing course in engineering and social science with several new courses in business administration. . . . An unhappy alumnus recalled that Course XV was "really a course in scientific management and Frederick Taylorism." (84-86)

Appendix 4D | Industrial Engineering Faculty

Abhyankar, Shreeram 1963-
Adams, Richard F. 1964-70
Amrine, Harold T. 1946-81
Anderson, James G. 1970-73
Baker, Norman R. 1965-69
Balyeat, Ralph E. 1949-56
†Barany, James W. 1958-
Barash, Moshe M. 1963-93
Bartlet, Thomas E. 1956-63
Beese, Charles 1937-46
Benjamin, Charles 1909-21
Bolz, Harold 1946-54
Brooks, George H. 1959-63
†Buck, James R. 1965-81
Caldwell, Barrett 2001-
Chandrasekar, Srinivasan 1986-
Chandru, Vijaya 1982-93
Chang, Tien-Chien 1982-2006
Charnes, Abraham 1955-57
Cheng, Gary 2007-
Chu, Chong Nam 1986-2002
‡Compton, W. Dale 1988-2004
Coullard, Collette R. 1985-90
Davis, Robert D. 1963-68
Deisenroth, Michael P. 1973-80
Derjani, Antoinette B. 1999
Dolan, Joseph B. 1982-94
Duffy, Vincent 2005-
†Eberts, Ray E. 1983-2003
†ElGomayel, Joseph I. 1967-2000
Engi, Dennis 2000-05
Field, Robert W. 1942-50
Feyen, Robert 2001-
†Fraser, Jane M. 1981-86
Gambrell, C. B. Jr. 1956-58
Gilbreth, Lillian 1935-48

Golden, Michael 1884-1916
Goss, William F. M. 1879-1907
Govinderaj, T. 1979-82
Greene, James H. 1948-85
Hartje, G. Fred 1955-64
Healy, Kevin 1992-97
Herman, Englebert W. 1955-61
Hicks, Charles H. 1957-61
Hill, Thomas W. Jr. 1967-73
Hockema, Frank C. 1930-56
Hoffman, James 1890-1938
Hulley, Oliver 1946-54
Johnson, William 1984-89
Kirkpatrick, Elwood G. 1946-79
†Koubek, Richard 1991-97
Landry, Steven 2005-
†Lascoe, Orville D. 1942-78
Lawley, Mark A. 1997-
Lee, Seokcheon 2007-
Lehto, Mark R. 1986-
Leimkuhler, Ferdinand 1961-99
Lindley, Roy 1921-63
Liu, C. Richard 1978-
Mann, Stuart H. 1970-72
McDowell, Richard W. 1955-62
Meier, Wilbur L. 1974-81
Montreuil, Benoit 1986-88
†Moodie, Colin L. 1964-99
Morin, Thomas L. 1975-
Mundel, Marvin 1942-52
Muthuraman, Kumar 2003-
Nof, Shimon Y. 1977-
O'Cinneide, Colm A. 1991-2001
Olson, David G. 1971-73
Owen, Halsey F. 1938-63
Ozsen, Leyla 2004-

† Indicates a professor voted *Best Teacher* by the students
‡ Indicates a professor recognized as the James H. Greene Outstanding Graduate Educator

†Papastavrou, Jason 1990-2000
Pekny, Joseph 1989-
Petersen, Clifford C. 1971-86
Phillips, Don T. 1971-75
Pigage, Leo 1941-47
Potter, A. A. 1920-53
Prabhu, Nagabhushana 1991-
Pritsker, A. Alan B. 1970-98
Radkins, Andrew P. 1954-58
Randolph, Paul H. 1957-61
†Rardin, Ronald L. 1982-2005
†Ravindran, A. 1969-87
Reed, Phillip A. 1968-72
Reed, Ruddell Jr. 1963-77
Reynolds, Gary 1970-72
Richard, Jean-Philippe 2002-
Richardson, W. 1949-52
Ritchey, John A. 1955-68, 85-90
Roberts, Stephen D. 1972-90
Rogers, H. Barrett 1938-42
Ryan, Jennifer 1997-2004
Sadowski, Randall P. 1976-88
Salvendy, Gavriel 1971-2008
†‡Schmeiser, Bruce W. 1979-
Shaw, Dong 1992-99
Shepard, George H. 1919-39
Smith, Clyde P. 1950-64
Solberg, James J. 1971-2006
Sparrow, Frederick T. 1978-2007
†Stewart, William T. 1980-82
Stojnic, Mihailo 2007-
Stuart, Julie Ann 2000-06
Sweet, Arnold L. 1973-2006
Taaffe, Michael R. 1981-89
Talavage, Joseph J. 1973-2000
Tanchoco, Jose M. A. 1984-

‡Thomas, Marlin U. 1993-2006
Tilles, Seymour 1950-52
Tu, Jay 1992-2003
Turner, William 1886-1937
†‡Uzsoy, Reha 1990-
Uhan, Nelson 2008-
Vellinger, Anthony J. 1915-63
Wagner, Donald K. 1982-90
Wallace, Lawrence 1906-13
Walters, Jack 1926-41
Wan, Hong 2004-
Wilson, James R. 1985-91
Yi, Ji Soo 2008-
†Yih, Yuehwern 1989-
Young, Hewitt H. 1955-67

Appendix 4E | Faculty Portraits

Early Engineering Deans

William Goss   Charles Benjamin   A. A. Potter   George Hawkins

Industrial Engineering Faculty before 1940

Charles Beese   Lillian Gilbreth   Michael Golden   Frank Hockema

James Hoffman   George Shepard   William Turner   Lawrence Wallace

Industrial Engineering Faculty 1940 - 2008

Harold Amrine   James Barany   Moshe Barash   James Buck

Barrett Caldwell

S. Chandrasekar

Vijaya Chandru

Tien-Chien Chang

Gary Cheng

W. Dale Compton

Collette Coullard

Michael Deisenroth

Vincent Duffy

Ray Eberts

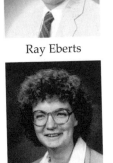
Joseph ElGomayel

Dennis Engi

Robert Feyen

Jane Fraser

James Greene

Richard Koubek

Stephen Landry

Orville Lascoe

Mark Lawley

Seokcheon Lee

Mark Lehto   Ferd Leimkuhler   C. Richard Liu   Wilbur Meier

Colin Moodie   Thomas Morin   Marvin Mundel   K. Muthuraman

Shimon Nof   Leyla Ozsen   Joseph Pekny   N. Prabhu

Clifford Petersen   Alan Pritsker   Ronald Rardin   Ravi Ravindran

Ruddell Reed   J.-P. Richard   Stephen Roberts   Randall Sadowski

Gavriel Salvendy   Bruce Schmeiser   James Solberg   F. T. Sparrow

Mihailo Stojnic   Julie Stuart   Arnold Sweet   Michael Taaffe

Joseph Talavage   Jose Tanchoco   Marlin Thomas   Jay Tu

Nelson Uhan   Reha Uzsoy   Donald Wagner   Hong Wan

James Wilson   Ji Soo Yi   Yuehwern Yih   Hewitt Young

Notes

1. Towne, "Engineer as an Economist," 428.
2. Taylor, *Scientific Management*, 1:6.
3. Qtd. in Wrege and Greenwood, *Frederick W. Taylor*, 217.
4. Nelson, *Mental Revolution*, 77.
5. Diemer, *Factory Organization*, 6.
6. Purdue University, *Fortieth Annual Catalog*, 145, 151.
7. Hannum, "Industrial Engineering," 68.
8. Knoll, *Story of Purdue Engineering*, 111.
9. Eckles, *Dean*, 14.
10. Knoll, *Story of Purdue Engineering*, 409.
11. Purdue University, *Sixty-Fourth Annual Catalog*, 125.
12. Shepard, *Elements of Industrial Engineering*, 28.
13. Ibid.
14. Ibid., 4.
15. Ibid., 247.
16. Knoll, *Story of Purdue Engineering*, 181.
17. Ibid., 180.
18. Ibid.
19. Drucker, *Post-Capitalist Society*, 36.
20. Time, "Purdue's Rocket Man," 52.
21. Hawkins, *Engineering Education*, 1.
22. DeGarmo, "Industrial Engineering," 70.
23. Hawkins, *Engineering Education*, 9.

5 | Operations Research

The resurgence of industrial engineering after World War II was propelled by the development of a field of study called operations research (OR) that emerged from the war as a new method of studying industrial problems in a comprehensive and rigorous way. OR had a radical influence on the traditional areas of manufacturing and work design through its intensive use of probability, statistics, applied mathematics, and computer science. It became the dominant area of graduate study and research for postwar industrial engineering departments.

In England during World War II, the name, *operational research* was used to describe, in a deliberately vague way, the work of top British scientists who were mobilized to give advice on military operations. OR teams were formed to study a wide variety of logistical problems; this practice was imitated in America with the formation of Navy, Army, and Air Force groups, where the name was changed to *operations research*. Sir Charles Goodeve described it as the application of science to executive problems, and gave an example of work done by the Royal Air Force group.

> In 1940 we had relatively few fighter aircraft and the most important new feature, radar, required mathematical calculations beyond the experience of ordinary commanding officers. Accordingly, a small party of half a dozen scientists was attached to Fighter Command and their analysis formed the basis for the operation of the whole defense organization of Britain. It is estimated that radar itself increased the probability of interception by a factor of ten; but that, in addition, this small operational research team increased the probability by a further factor of about two, which together meant that the Air Force was made twenty times more powerful.[1]

In an excellent history, *Operational Research in War and Peace*, Maurice Kirby traces the linkage between OR and scientific management.

117

Taylor's approach to managerial organization can be regarded as a direct precursor of operational research because it was concerned to optimize the use of existing resources by reference to procedures akin to the "scientific method," based upon observation and measurement . . . although it is important to note that its techniques of analysis were far removed from the relative technical rigor of operational research, as it was to develop after 1950. The fact remains that scientific management originated in *practical* experience on the shop floor where "intuitive feel" had an important role to play in achieving greater operational efficiency.[2]

Kirby noted that the first school in England to adopt OR was the Institute for Engineering Production at the University of Birmingham in 1958. There was a similar pattern of development in America after the war, beginning at several universities where faculty and students conducted OR studies for corporate and military sponsors. At Johns Hopkins University, home of the Army's OR center, Dean Robert Roy started a new graduate program in industrial engineering in 1955 with OR as its base. Courses in mathematics, statistics, economics, and psychology were combined with application courses in stochastics, optimization, game theory, and inventory. The plan suited the postwar need of the field of industrial engineering for graduate study and it was adopted quickly at other IE schools.

There was much debate over the theoretical and pragmatic nature of OR and whether it was science, engineering, or management. Only recently, in 1995, did the two leading U.S. societies merge to form the International Federation of Operations Research and Management Science (INFORMS). Philosopher C. West Churchman, an early OR society president, thought OR was a "systems approach" to solving the large societal problems that require optimal solutions—not the suboptimal, "one best way" solutions of Taylor and the Gilbreths.

If we fail to think about the larger system, our thinking becomes fallacious. It is necessary to make strong assumptions about the boundary between a problem and its environment in order to balance the local objectives with global ones. . . . Suppose Henry Ford had asked: "Should I, morally, make cheap autos?" He might very well have replied yes and that his contemporary automakers were immoral because they built cars only for the rich. However, a time traveler would have been able to tell Henry that his cheap autos would eventually produce the smoggy cities of today.[3]

On the other hand, Herbert Simon thought Babbage and Taylor should be made retroactive members of the operations research societies.

No meaningful line can be drawn any more to demarcate operations research from scientific management or scientific management from management science. Along with some mathematical tools, OR brought into management decision making a point of view called the systems approach that is no easier to define than OR, for it is a set of attitudes and a frame of mind rather than a definite and explicit theory. It means looking at the whole problem—again, hardly a novel idea, and not always a very helpful one. Somewhat more concretely, it means designing the components of a system.[4]

Philip Morse, head of MIT's physics department and first president of the Operations Research Society of America, stressed the need for "hands on" learning.

OR is a scientific study in its own right, using experimental as well as theoretical techniques to study a natural phenomenon, an operation. Its object is to *understand* the behavior of various operations, so as to *predict* the operational result of changes in operating rules or equipment, and thus to better *control* the operation and improve its result. . . . It is futile to try to carry on operations research without detailed observations of the operations and without close contact with the staff managing it, and above all the operations worker must be able to work with people and appreciate the non-scientific aspects of the management problems he is helping to solve. . . . Computations and equations put down on paper don't upset the boss; but it takes a lot of argument to persuade him to monkey with the production line.[5]

At Hopkins, Roy was concerned about the emphasis on analysis over design. "For the OR pioneers, the problem was the thing—destroy U-boats, intercept bombers," Roy said, "not derive universal models."

Within the population of those identified with OR there has come to be a bimodal, almost bipolar kind of distribution, and the two parts do not always meet in mutual esteem. There are those whose talents and interests are in *operations research* and those whose talents and interests are in *operations engineering*, "discoverers" in contrast to "doers." When the goals of an organization are to improve the performance of operations, the need is for operations engineers. When the goals of an organization are to create new concepts and methods, the need is for operations researchers. . . . The investigator who discovers naught save trivia is likely to be more snob than creator. The applier who becomes an obsolete "handbook" practitioner is likely to be more hack than professional.[6]

Amid the discussion that still goes on today, the academic discipline of operations research focused on "how to form mathematical models of complex engineering and management problems and how to analyze them to gain insight about possible solutions," to quote Purdue professor Ronald Rardin in his recent book, *Optimization in Operations Research.*[7] Another Purdue professor, James Solberg, points out the value of mathematical models in enabling one to see problems differently and more clearly.

> For all of our modern technology and sophistication, most people today are quite poor decision makers, particularly in the face of uncertainty. Psychologists who study human decision making point out that irrational behavior is so prevalent that one wonders how our species could have survived to now. What is called "common sense" is so often wrong that it should be regarded as dangerously misleading.[8]

Solberg describes the art of mathematical modeling in Appendix 5A and in his award-winning textbook, *Operations Research: Principles and Practice*, written with Purdue professors Ravindran and Phillips.[9]

OR modeling is divided into two major areas: optimization and stochastic models. Optimization models are concerned with the allocation of scarce resources: labor, materials, machines, and capital in the best way so as to minimize costs or maximize benefits. The general method of solving such problems is called *programming*, and the most famous example is that of George Dantzig, who developed a powerful solution technique called the *simplex method*. His example problem of assigning seventy people to do seventy jobs seems to be unbelievably difficult because the number of possible assignments is greater than the number of particles in the universe. However, Dantzig showed that an optimal assignment could be found quickly using his simplex algorithm.

Optimal allocation of resources has been the objective of engineering economics problems since they were first formulated. The earliest textbook on engineering economy was Wellington's *The Eco-*

nomic Theory of Railway Location of 1887, in which he said: "Engineering is the art of doing well for one dollar what any bungler could do with two."[10] Engineering economy, the study of investments in technical infrastructures, has been and still is a foundational topic in industrial engineering. The subject was well developed in 1930 as seen in classic textbooks like Eugene Grant's book, *Principles of Engineering Economy*, that has gone through ten editions.

In addition to Dantzig, pioneers in mathematical programming included Bellman, Charnes, Koopmans, and von Neumann. Important extensions have been made in areas like inventory management by Whitten, network analysis by Ford and Fulkerson, and scheduling by Conway, Maxwell and Miller. Recently, former Purdue IE professor Vijay Chandru applied it to logical inference problems in artificial intelligence, computer science, and manufacturing.[11]

The second major kind of operations research model deals with the design of stochastic systems that exhibit changes that are unforeseen and difficult to predict. Real-world examples are traffic jams on busy highways and waiting lines at service centers of all kinds. Stochastic analysis, or queueing theory, is a statistical way of studying such problems, that began with the efforts of a telephone engineer, A. K. Erlang, to estimate the average number of customers that would wait for service under given conditions. His model was used widely in the 1920s and Philip Morse at MIT drew on this theory to study submarine warfare in World War II.

The theory of stochastic modeling and all of statistical analysis rests on the probability theory that came from early studies of gambling. Around 1800, the work of Gauss, Laplace, Quetelet, and Galton led to reliable methods of statistical estimation that were perfected in the early 1900s by Cochran, Fisher, Neyman, Pearson, and Yule. Pioneering work in biostatistics was done at that time by Florence Nightingale who invented the pie chart. Important applications of statistics to industry were made at Bell Laboratories in the 1920s and 1930s by Shewhart, who invented the quality control chart, and Dodge and Romig, who developed inspection plans. Industry was slow to adopt quality control until World War II, when the War Department set standards and started training programs.

Quality control was a key factor in the industrial recovery of Japanese manufacturing after the war through the efforts of Deming and Juran. Deming, who was a mathematician of note, is better known for the practical QC program shown in Table 5A, which seems to be an extension of the scientific management philosophy of Frederick Taylor and the Gilbreths. The Deming Prize is the most prestigious

Table 5A. *Deming's fourteen point program for quality control* [12]

1. Continually improve products and service to society.
2. Eliminate defective products and worker incompetence.
3. Eliminate mass inspection. Build a quality product and process.
4. Choose suppliers based on quality not price.
5. Constantly improve planning, production, and service.
6. All employees must be properly trained and up-to-date.
7. Managers help employees do a better job.
8. Employees must feel secure and unafraid to report problems.
9. Break down barriers between departments.
10. Eliminate slogans and let workers set realistic goals.
11. Eliminate quotas that make workers ignore quality.
12. Listen to worker suggestions and encourage pride in work.
13. Educate managers and workers in teamwork and QC techniques.
14. Top management must push the first thirteen points every day.

industrial award in Japan. Juran's adage that "quality must grow from the bottom up and be fostered from the top down" could have been a quote from Lillian Gilbreth.

The first Ph.D. in IE at Purdue was awarded in 1942 to Albert Johnson for a statistical analysis of admissions. The first professor to teach operations research courses at Purdue was E. "Kirk" Kirkpatrick, who joined the faculty in 1946 from Case Institute and taught courses in engineering economy, statistics, and quality control. Charles Hicks and Paul Randolph taught statistics and stochastic theory in the 1950s, and, when they transferred their affiliation to the newly formed department of statistics in the 1960s, Prof. James Barany took the lead in teaching basic statistics at the undergraduate level and experimental design and analysis at the graduate level for over fifty years. Over the years there has been a close connection between IE and the statistics department started by Profs. Hicks and Virgil Anderson.

I joined the faculty in 1961 and taught courses in engineering economy and stochastic processes. A large grant I received from the National Science Foundation for an OR study of research libraries was part of a larger NSF program to improve science communication involving OR leaders like Churchman at Case, Charnes at Texas, Roy at Hopkins, and Morse at MIT.[13] As at MIT, Purdue libraries became an OR teaching laboratory with the support of its director Jack Moriarty for the work of Profs. Norman Baker, Thomas Hill and I, along with our many students.

Teaching of stochastic processes at Purdue changed markedly with the arrival of Profs. James Anderson, Clifford Petersen, Alan Pritsker, James Solberg, Arnold Sweet, and Joseph Talavage, who were collaborators in the Large Scale Systems Research Center organized by Pritsker in 1970 to study complex systems using computer simulation as discussed in chapter eight. Much of their research contributed to the development of computer simulation, which is tied closely to the theory of stochastic processes as reflected in the Ph.D. research done by the students of Solberg and Sweet and also those of Bruce Schmeiser, Michael Taaffe, and James Wilson who joined the faculty in the 1980s.

Solberg's model of automated manufacturing systems as a queueing network (CAMQ) was widely used as a design tool. Teaching and research in stochastic analysis expanded in the 1990s and later with the addition of faculty members Dennis Engi, Colm O'Cinneide, Jason Papastavrou, Marlin Thomas, and Hong Wan. Research accomplishments are seen in the Ph.D. theses listed in Appendix 10A. Both Engi and Thomas were school heads during their stays on the faculty.

Optimization modeling was first taught at Purdue in the 1950s by Abraham Charnes and Thomas Bartlet. Charnes was a prominent figure in the early research done in the area. He came from Carnegie Mellon before going on to Northwestern and then Texas. Bartlet came from Hughes Aircraft coincident with Charnes but stayed longer, returning to California in 1963 for consulting work in hospital planning. The courses they launched, IE 335, 535, and 635, are still being taught over fifty years later (see Appendices 5C and 9A).

The optimization faculty was enlarged in the 1960s with the arrival of five professors. Two were Charnes's students from Northwestern, Norman Baker and Robert Davis, and two were George Dantzig's students from Berkeley, Gary Reynolds and Ravi Ravindran. The fifth was Thomas Hill, Alan Pritsker's student from Arizona State. Baker went on to Cincinnati where he became chancellor, and Ravindran became a vice president at Oklahoma and then head of the IE program at Penn State.

A second surge in optimization research began in the 1970s with the arrival of Wilbur Meier, Thomas Morin, Don Phillips, and Frederick Sparrow, followed by Vijay Chandru, Collette R. Coullard, Jane M. Fraser, Rardin, and Don Wagner in the 1980s. More recent appointments in the optimization area were those of Nagabhushana Prabhu, Kumar Muthuraman, Jean-Philippe Richard, Dong Shaw, Mihailo Stojnic and Nelson Uhan. This faculty team was augmented by the courtesy appointment of Shreeram Abhyankar, Distinguished Marshall Professor of Mathematics.

In 1986 Tom Morin was principal investigator for a $4.25 million grant from the Department of Defense for research on "computational combinatorics" with Profs. Abhyankar, Chandru, Coullard, Rardin, and Wagner, in collaboration with faculty and students in the Department of Computer Science and the School of Management. O'Cinneide, Papastavrou and Shaw left Purdue for Wall Street where there was increasing interest in applying operations research to financial markets.

Sparrow's students modeled large scale transportation and energy systems with support from the Purdue Institute for Interdisciplinary Engineering Studies (IIES) and the Indiana State Utility Forecasting Group (SUFG), both of which Sparrow headed in the 1980s and 1990s. IIES funded studies on transportation mobility and renewable energy. Sparrow's SUFG team of analysts have been credited with preventing Indiana and others from joining the rush to deregulation that, for example, became such an enormously costly mistake for the State of California in 2000.

The SUFG team used OR game models of market failure in electricity and gas markets to show how deregulation might simply replace a regulated monopoly with an unregulated monopoly or duopoly that would raise rather than lower costs. Their work in predicting an "artificial scarcity" in supposedly open electricity markets awakened regulators to the gaming opportunities inherent in a deregulated electricity and gas industry. The SUFG study of *Demand Side Management* (DSM) electric power "conservation investments" warned regulators of flaws in the program that would end up costing ratepayers far more than any savings. As a result, Indiana and other states curtailed their involvement and were spared the cost and inefficiency of the now abandoned DSM programs. They also showed that open competition in the electricity industry could increase, rather than decrease, prices in Indiana and Kentucky.

The standout application of operations research at Purdue was its use in the huge CIDMAC/ERC study of computer integrated manufacturing systems led by Solberg. Eighty professors and a thousand students were involved in the effort that lasted twenty years at a cost of almost one hundred million dollars in federal and corporate grants. It was the largest research project in Purdue's history and had a major impact on the direction taken by the U.S. manufacturing industry in response to global competition. The CIDMAC/ERC project is described in detail in chapter 6.

In 1981 Russell Ackoff, a close colleague of Churchman and also an early ORSA president, criticized the operations researchers' approach to systems planning for their reliance on analytic models.

> There are three kinds of things that can be done about problems—they can be *resolved, solved,* or *dissolved.* To *resolve* a problem is to select a course of action that is good enough, that *satisfices* (satisfies and suffices). We call this approach *clinical* because it relies heavily on past experience [and] most managers are problem resolvers [because it] minimizes risk and therefore maximizes the likelihood of survival.
>
> To *solve* a problem is to select a course of action that is believed to yield the *best* possible outcome that *optimizes.* We call this the *research* approach because it is largely based on scientific methods, techniques, and tools.
>
> To *dissolve* a problem is to change the nature and/or environment of the entity in which it is embedded so as to remove the problem. Problem dissolvers *idealize* rather than satisfice or optimize [and] we call this the *design* approach [where the] objective is *development* rather than survival or growth. . . . To develop is to increase one's ability and desire to improve one's own quality of life and that of others.[14]

Ackoff said design-oriented planners must be competent in the use of analytic tools and must be able to draw on clinical experience, but they must be generalists—humanists as well as scientists, at home as much with art as with technology—taking a holistic approach. By holistic, he meant they must first have a conception of the whole of the system before designing the parts. An analytic or reductionist approach, on the other hand, tends to design the whole after designing the parts. He said a car disassembled into its parts is no longer a car, and one cannot realize a car by first designing its parts and then assembling them into a whole.

Ackoff's challenge corresponded to the theme of alumnus Gerald Nadler's treatise, *Breakthrough Thinking: Why We Must Change the Way We Solve Problems, and the Seven Principles to Achieve This.* His principles, summarized in Appendix 5B, focused on the design of a system with a unique purpose and a clear understanding of how the system being designed relates to the larger system in which it is embedded as well as the subsystems it contains. Nadler was also concerned with finding the information needed in a participative, dynamic, and future-oriented way.

The overall design concept has to be big enough to deal with all of the important qualitative and quantitative requirements of the situation. If the initial attempt fails to make a real difference, than a still more demanding concept has to be found. It is then that the resources of specialized knowledge are brought into play through the coordination of many different specialties. The leadership must communicate a crystal clear vision of the overall objective as well as the subsidiary objectives. Such an operations research enterprise must be aimed at making a real difference and recognize the fact that success will require hard work, not just flashes of genius.

Appendix 5A | The Art of Modeling

The following excerpt is from Prof. James Solberg's class notes on "Stochastic Modeling for Industrial Engineers" in his book, Stochastic Modeling.

The first step is construction of the model itself, which is indicated by the line labeled Formulation (in Figure 5A). This step requires a set of coordinated decisions as to what aspects of the real system should be incorporated in the model, what aspects can be ignored, what assumptions can and should be made, into what form the model should be cast, and so on. In some instances, formulation may require no particular creative skill, but in most cases—certainly the interesting ones—it is decidedly an art. The selection of the essential attributes of the real system and the omission of the irrelevant ones require a kind of selective perception that cannot be defined by any precise algorithm.

Too often in practice, the formulation step is made unconsciously. That is, a person who is familiar with one class of models tends to view reality in those terms and consequently recognize only those aspects of reality that fit the class. Someone captured this myopia nicely with the phrase, "To he who has only a hammer, everything looks like a nail." A good analyst will have at his or her disposal a large repertoire of modeling tools, so that the choice can be influenced more by the nature of the problem than the limited capabilities of the modeler.

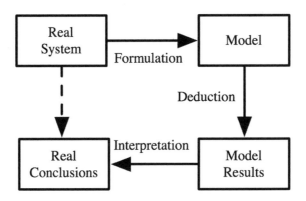

Figure 5A. *Solberg's diagram of the OR method of modeling reality*

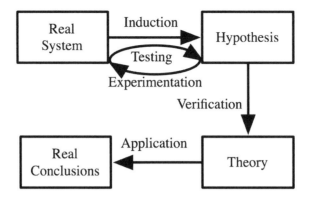

Figure 5B. *Solberg's diagram of the scientific method*

Once the problem formulation and definition is agreed upon, a more rigorous step in the modeling process is begun. Referring (to Figure 5A), the step labeled Deduction involves techniques that depend on the nature of the model and questions you want to answer. It may involve solving equations, running a computer program, expressing a sequence of logical statements—whatever is necessary to solve the mathematical problem. The final step, labeled Interpretation, again involves human judgment. The model conclusions must be translated to real-world conclusions cautiously, in full recognition of possible discrepancies between the model and its real-world referent.

The most important point revealed (by Figure 5A) is that the ties between the model and the system are at best ties of plausible association, and that no one, no matter how competent can achieve perfect association. . . . Despite these imperfections, a good model will provide much better insight into real-world situations than most people can achieve without the model. The process of acquiring the conviction that a model actually "works" is commonly called validation.

To further clarify the modeling approach to problem solving, contrast it to the experimentally based "scientific method" of the natural sciences. (Figure 5B) depicts the latter approach. Here, the first step is the development of a hypothesis, which is arrived at generally by induction, following a period of informational observation. At that point an experiment is devised to test the hypothesis. If the experimental results contradict the hypothesis, the hypothesis is revised and retested. The cycle continues until a verified hypothesis, or theory, is obtained. Theories are discovered; models are invented. Scientists want to understand the world; engineers want to improve it. (5-7)

Appendix 5B | Breakthrough Thinking

This is an abridgement taken from the book Breakthrough Thinking *by Gerald Nadler and Shozo Hibino.*

The purpose and spirit of Breakthrough Thinking is in no way exclusive or competitive. We intend to demonstrate that all humanity needs to help solving its problems and that sharing knowledge about *how* to do this is good for all of us. Ultimately, individual human beings and how they think are important. To transform individual lives, leaders of organizations must themselves change and demonstrate a new way of thinking.

As one group or nation becomes increasingly able to find better solutions so to will other groups and nations be able to find other sets of good solutions. All nations should be looking to the future. And in a world increasingly independent, the truly effective solution of one nation's problem will contribute to the betterment of all.

What is essential now is that you mentally begin to make the fundamental shift necessary to learn and apply the process of Breakthrough Thinking. . . . The gap each of us must overcome needs another concept—a set of principles or assumptions and a process of reasoning to put the principles into action. Here, then, ends your introduction to the seven principles.

1. *The Uniqueness Principle*: Each problem is unique and may require a unique solution.

2. *The Purposes Principle*: Focusing on and expanding your purposes help strip away nonessential aspects of a problem.

3. *The Systems Principle*: Every problem is part of a larger system of problems, and solving one problem inevitably leads to another. Having a clear framework of what elements and dimensions comprise a solution insures its workability and implementation.

4. *The Solution-After-Next-Principle*: having a target solution in the future gives direction to near-term solutions and infuses them with large purposes.

5. *The Limited Information Collection Principle*: Excessive data gathering may create an expert in the problem area, but knowing too

much about it will probably prevent discovery of some excellent alternatives.

6. *The People Design Principle*: Those who will carry out and use the solution should be intimately and continuously involved in its development. Also, in designing for the other people, the solution should include the critical details to allow some flexibility to those who must apply the solution.

7. *The Betterment Timeline Principle.* The only way to preserve the vitality of a solution is to build in and then monitor a program of continual change. The sequence of Breakthrough thinking solutions thus becomes bridge to a better future. (88-89)

Appendix 5C | Graduate Courses in OR

IE 501 Introduction to Operations Research. Fundamentals of operations research. Mathematical programming, decision theory, stochastic processes, and their applications. Emphasis on problem formulation, solution strategies, and computer software packages.

IE 530 Quality Control. Principles and practices of statistical quality control in industry. Control charts for measurements and for attributes. Acceptance sampling by attributes and by measurements. Standard sampling plans. Sequential analysis. Sampling inspection of continuous production.

IE 532 Reliability. Reliability of components and multi-component systems. Application of quantitative methods to the design and evaluation of engineering and industrial systems and processes for assuring reliability of performance. Economic and manufacturing control activities related to product engineering aspects of reliability. Principles of maintainability. Product failure and legal liability.

IE 533 Industrial Applications of Statistics. The application of statistics to the effective design and analysis of industrial studies relating to manufacturing and human factors engineering in order to optimize the utilization of equipment and resources. Emphasis on conducting these studies at the least cost.

IE 535 Linear Programming. Optimization of linear objective functions subject to linear constraints. Development of theory and algorithmic strategies for solving linear programming problems.

IE 536 Stochastic Models In Operations Research I. An introduction to techniques for modeling random processes used in operations research. Markov chains, continuous time Markov processes, Markovian queues, reliability and inventory models.

IE 537 Discrete Optimization Models and Algorithms. An introduction to classic models and algorithms for discrete optimization. Basic theory and computational strategies for exact and heuristic solution of integer, combinational, and network problems.

IE 538 Nonlinear Optimization Algorithms And Models. Survey of computational tools for solving constrained and unconstrained nonlinear optimization problems. Emphasis on algorithmic strategies and characteristic structures of nonlinear problems.

IE 539 Stochastic Service Systems. Theory and application of models of stochastic service systems. Stationary Markov models, nonstationary Markov models, and general models. Numerical algorithms and approximation methods are emphasized.

IE 545 Engineering Economic Analysis. Analysis of engineering costs and capital investments. Applications of classical optimization, mathematical programming, and the theory of production to the analysis of investment proposals. Evaluation and selection of individual projects and formulation of capital investment programs.

IE 546 Economic Decisions in Engineering. Topics in decision making and rationality including decision analysis, decision making under uncertainty, and various descriptive and prescriptive models from operations research, economics, psychology, and business. Applications are drawn from engineering decision making, public policy, and personal decision making. Attention also is paid to designing aids to improve decision making.

IE 630 Multi-Objective Optimization. Theory and applications of multiple objective optimization and multiple criteria decision making. Existence and specification of single and multiple attributive value and utility functions. Risk aversion. Characterization and generation of efficient (Pareto optimal) solutions to vector optimization problems. Domination cones. Interactive algorithms.

IE 631 Heuristic Optimization. Theory and applications of methodological approaches explicitly addressed to heuristic or approximate optimization of mathematical (primarily discrete) models. Worst-case/ expected-case/empirical approaches to evaluating heuristic schemes, generic greedy, truncated exponential and local search algorithms, and approaches not usually considered mathematical programming, such as artificial intelligence search and interactive optimization. Emphasis on viewing available results as generic methodology that can be studied in the same algorithmic framework as exact methods.

IE 632 Scheduling Models. Development and discussion of mathematical and simulation models for scheduling and sequencing the flow of jobs or activities in manufacturing environments. Techniques include dynamic programming, branch and bound, linear and integer programming, heuristics and stochastic network simulation.

IE 633 Dynamic Programming. Theory and applications of finite and infinite stage sequential decision processes, including Markovian decision problems, efficient computational methods, and computer software package.

IE 634 Integer Programming. An advanced course on theory and algorithms for integer and mixed integer optimization problems. Convergence of integer programming algorithms, dual relaxations, Benders decomposition, cutting plane theory, group theory of integer programs, and linear diophantine equations.

IE 635 Theoretic Foundations of Optimization. An advanced course in theoretical foundations of mathematical programming. Convex analysis, global and local duality and optimality, general algorithmic convergence.

IE 636 Stochastic Models in Operations Research II. The continued development of mathematical models of discrete valued stochastic phenomena, with applications to queueing processes, machine repair problems and inventory policy. Poisson processes, renewal processes, semi-Markov and Markov renewal processes. Regenerative processes and renewal reward processes. Multi-dimensional birth and death queueing models.

IE 637 Computational Methods for Queueing Networks. Study of computational methods for stochastic models, especially models of manufacturing, traffic, and communication systems. Topics include: decomposition methods, methods based on Markov chains, transform inversion, heavy-traffic approximations, and algorithms for product-form networks.

IE 638 Engineering Technical Forecasting Methods. Development of techniques used for forecasting time series. Introduction to time series, covariance functions, solution of difference equations, autoregressive, moving average, and seasonal forecasting. Use of the Box and Jenkins computer programs. Exponential smoothing and other ad hoc models. Applications to data gathered from various industries. Technological forecasting, analogy models, growth models, trend extrapolation, breakthroughs, and other methodologies.

IE 639 Combinatorial Optimization. An advanced course in combinatorial optimization. Emphasis on the theory and strategies of combinatorial algorithms including complexity theory, advance issues in network flows, matching, matroids, and combinatorial polyhedra.

Notes

1. Goodeve, "Operational Research," 380.
2. Kirby, *Operational Research*, 50.
3. Churchman, *Systems Approach*, 125.
4. Simon, *New Science*, 14.
5. Morse, *Notes on Operations Research*, 1.
6. Roy, *Cultures of Management*, 284.
7. Rardin, *Optimization in Operations Research*, 1.
8. Solberg, *Stochastic Modeling*, 3.
9. Ravindran, Phillips, and Solberg, *Operations Research*.
10. Wellington, *Economic Theory*, 1.
11. Chandru and Hooker, *Optimization Methods*.
12. Deming, *Out of the Crisis*, 23.
13. Swanson, *Operations Research*, 84.
14. Ackoff, "The Art and Science," 20-21.

6 | Manufacturing

Industrial engineering can trace its development to early man's efforts to make tools and use them to satisfy basic needs. Charles Babbage thought it wasn't until the eighteenth century that civilization reached a level where it could organize the kind of mass-production in factories that emerged in the Industrial Revolution. Factory tooling reached a very high level in the eighteenth and nineteenth centuries, especially in textiles and metals. Lathes go back to ancient times, when they were first powered by bent trees, later by water in the Middle Ages, and then by steam in the Industrial Revolution. Maudslay's screw lathe, introduced about 1800, astounded his contemporaries for the workmanship and accuracy of the parts it produced. It was a key building block in the English system of precision manufacturing and enabled Britain to become the dominant industrial nation that it celebrated with the Great Crystal Palace Exhibition of 1851 in London.

By the end of the nineteenth century, however, the English system of manufacturing was supplanted by the American mass-production system that was pioneered by Eli Whitney. Faced with a lack of access to English craftsmanship and precision tooling and the need to fulfill a contract to make muskets for the U.S. Treasury, Whitney devised a way to emulate the work of master smiths with a series of tasks that could be executed by ordinary workmen using specially designed machine tools. Rather than aim for "work-of-art" precision, Whitney determined the amount of deviation that could be tolerated in the components, and made "go-no-go" gauges to maintain satisfactory accuracy.

The muskets Whitney assembled were functionally equivalent to the handcrafted work of a master gunsmith and they could be repaired quickly in the field, an important advantage of interchangeable parts. Whitney invented special tools for his system; in particular, his milling machine of 1818 is regarded as one of the great machine-tool inventions of all time, on par with Maudslay's lathe. Following its demonstration at the 1851 Exhibition in London, the Whitney system was adopted by the world's largest armories, Enfield in Britain, Beretta in Italy, and Colt in America. The Colt Armory employed 600 workmen and 400 machine tools powered by a single large steam engine to produce 250 arms per day.

Whitney's system became the model for American metal manufacturing of all kinds. Mass-produced plows were introduced in the 1850s, sewing machines in the 1860s, and typewriters in the 1880s. 400,000 U. S. bicycles were made in 1894. By 1900, the basic version of almost every machine tool used today had been invented. Automobile factories became spectacular examples of mass-production systems. Olds made 2,500 cars in 1902 and 5,000 in 1904. In 1914 Henry Ford's Highland Park plant produced a thousand Model-Ts per day with a workforce of 15,000. Ford assembly lines became the symbol of mass-production for a half century.

Textile production underwent changes that were as significant as those of metal manufacturing. Arkwright's cotton mill was the first factory in the Industrial Revolution. Norbert Wiener, famous for his work on feedback control, wrote about the evolution of the textile factory.

> The textile mills furnished the model for the whole course of the mechanization of industry. On the social side, they began the transfer of workers from the home to the factory and from the county to the city. . . . Many of the disastrous consequences and phases of the earlier part of the industrial revolution were not so much due to any moral obtuseness, but to technical features inherent in the early means of industrialization. It was necessary to collect the machinery in large factories, where many looms and spindles could be run from one steam engine. . . . Even as late as the time of my own childhood, the typical picture of a factory was that of a great shed with long lines of shafts suspended from the rafters, and pulleys connected by belts to individual machines.[1]

Wiener said that nobody foresaw how electricity would revolutionize manufacturing. The electric motor untied machines from the pulleys and belts of a central power plant. Factories became quiet, open buildings with logical layouts that expedited the flow of work, paving the way for the modern automated factory of computers and robots.

The invention of the Bessemer blast furnace in England in 1856 caused a major revolution in the use of steel in all of its many applications. It led to the building of steel towns like Bethlehem, Pennsylvania, enabling America to quickly become the world's leading steel producer. It was at Bethlehem Steel in 1899 that Frederick Taylor and Maunsel White completed experiments that led to the discovery of "high speed steel," which retained its hardness at high temperatures and greatly increased the speed of machining operations. It doubled the efficiency of steel mills but required the redesign of machine tools to handle the speeds and an army of specialists to perform more precise tasks like fixturing, material handling, tool sharpening, and belt tightening. Taylor and Henry Gantt worked on the optimum combinations of the cutting variables, and Carl Barth devised a slide rule for making the calculations.

Taylor's larger goal at Bethlehem Steel was to create a new kind of manufacturing system in which every operation, including all human tasks, was subjected to the same kind of intensive, scientific study that he had given to steel cutting. The Taylor system required a team of industrial engineers to continually study and manage every aspect of a factory. Henry Ford's 1913 automobile assembly line became the popular symbol of Taylor's approach, even though Ford strenuously denied Taylor had any involvement in designing his factories. The American automobile industry later pushed the assembly line idea to an extreme—mechanizing and automating peopleless systems that would have astounded Taylor.

After World War II, manufacturing underwent many major changes such as the use of *statistical process control* (SPC) by the War Department to cover military purchases. This focused attention on the problem of making products that met certain quality standards in a consistent manner. The quality control approach proved to be an important factor in the rise of the Japanese automobile industry that depended much more on the skill and cooperation of its workforce. Another very important postwar development was the *numerical control* (NC) of machining operations using prerecorded digital instructions. NC made it possible to store, revise, and transfer machining know-how systematically and electronically and to control quality more closely. Steady improvements in NC hardware, software, and computer linkages reduced its cost and accelerated its adoption as a standard shop floor practice.

Numerical control fostered the development of three major, computer-based, automation innovations: *flexible manufacturing systems* (FMS), *computer-aided design* (CAD) and *computer-aided process planning* (CAPP). FMS used computers to coordinate the flow of materials and tools among machining centers, test stations, and storage facilities. An FMS could make a variety of products in small batches, or an entire family of parts simultaneously on the same set of equipment, and do it in an automated, timely, and cost effective way. It was a major achievement in reaching the long-run goal of fully automating the manufacture of products in small batches and on-demand.

The simultaneous development of CAD and its companion, CAPP, were significant additions to computerized manufacturing. CAD revolutionized the product design process by making it possible to consider a very large number of design alternatives. With the advent of the Internet, design could be done collaboratively around the globe. For example, a design team in India could take over from an American team at the conclusion of their workday. In doing for process design what CAD did for product design, CAPP overcame the troubling gap between the creation of new product ideas in a design laboratory and their realization in a factory.

Before CAPP, product designers complained about "throwing designs over a factory wall." General Motors merged its design and manufacturing organizations in order to minimize this communication problem. The combination of CAD and CAPP with FMS into a fully computer-integrated manufacturing (CIM) environment was the technical solution. In principle, it allowed engineers to design virtual products and processes that could be simulated in a virtual factory before beginning real production operations. Furthermore, this knowledge could be developed, distributed, and used communally over a large network of design, production, and sales centers.

CIM made it possible to overcome a major limitation of mass-production, the lack of variety in its products, as reflected in the quip about Henry Ford saying he could provide any color the customer wanted as long as it was black. With CIM, it was possible to automate low-volume production with a relatively short lead-time and approach the sales ideal of customized production and on-demand response to one-of-a-kind requests. Furthermore, such responsiveness could be achieved in a global network of suppliers.

Before the first product is made on a large-scale automated manufacturing system, all of the essential components—materials, tools, fixtures, programs, schedules, people, etc.—have to be in position, and this creates a complex information management challenge. Automation has brought manufacturing full circle, back to the time when a single master craftsman was in control of all aspects of the process. In this sense, the modern CIM operator is more like an expert metal smith in a Maudslay-type precision shop than the unskilled operator in a Whitney-type factory.

At Purdue, manufacturing engineering was first studied in the practical mechanics laboratories that resembled factory shops. In addition to their use as teaching laboratories, the practical mechanics shops were used for testing processes and fabricating products. William Goss invented and produced the Purdue Lathe as a teaching tool for his classes. The lathe was so popular that it was produced and sold to secondary schools and other universities as a teaching device. Knoll writes, "Students helped manufacture the lathe in their regular course work, and so many orders were received that not all of them could be filled."[2]

Michael Golden and William Turner were the other members of Goss's original "MIT gang," and after Golden and Goss left, Turner was joined by James Hoffman, Roy Lindley, and Tony Vellinger. In the Practical Mechanics building, later called Michael Golden Hall, there were complete facilities for forging, welding, machining, and woodworking. In 1937, when practical mechanics became part of general engineering, Hoffman and Turner retired, leaving antiquated facilities, according to Knoll.

> The machines were still run by leather belts from overhead shafts, and illumination was supplied through the windows and skylights and by four naked 100-watt bulbs that hung down from the ceiling. A visitor could have regarded the shop as an interesting industrial museum or could have been dismayed at the thought that it was supposed to be a place for training engineers. In a long gamble every piece of equipment was sold, in the hope that perhaps modern equipment could later be secured from war surplus. The market was good because of wartime demand for almost any sort of machine and it was gratifying to find that machines a half century old, but in good condition because of Deacon Turner's loving care, brought considerably more than the original purchase price.[3]

After the war, Orville Lascoe refurbished Michael Golden Labs with millions of dollars worth of war surplus equipment he obtained by paying only the shipping charges. He installed a sophisticated metrology lab in the basement of MGL with tooling from the Naval Gun Factory in Washington. Lascoe was a popular teacher in the tradition of Michael Golden and known for the annual tool builder conferences he hosted in MGL. To help teach the basic undergraduate courses in manufacturing, Lascoe and Vellinger were joined by Richard Adams, Fred Hartje, Richard McDowell, and Clyde Smith.

Lascoe's Ph.D. students studied a wide variety of manufacturing processes including polymers and ceramics processing, extrusion, and electric-discharge machining. One of Lascoe's students, Joseph ElGomayel, joined the Purdue faculty in 1963 to teach manufacturing processes and process planning using group technology.

Lascoe recruited Moshe Barash in 1963 from the University of Manchester Institute of Science and Technology (UMIST) in England, where he was a well-known authority on manufacturing developments in Eastern Europe. Twenty years after he came to Purdue, Barash was joined by an early Manchester colleague, William Johnson. (See Johnson's comments about Barash and Manchester in Appendix 6A.) Johnson was a professor of mechanics at Cambridge University, a Fellow of the Royal Society, and a Fellow of the Royal Academy of Engineers. At Purdue, he was given the title United Technologies Professor of Engineering.

Barash established a solid research program in machining, metal forming, and tribology, but his more famous initiative was organizing an NSF study of computer-integrated manufacturing (CIM), as described in Appendix 6B. This very large and comprehensive program eventually required the cooperation of many different engineering specialists. It began with a pioneering operations research and systems engineering study by Barash, Colin Moodie, James Solberg, Joseph Talavage, and the author, of the revolutionary FMS systems being developed by Caterpillar, Sundstrand, and Ingersoll-Rand.

In the late 1970s Purdue President Arthur Hansen embarked on a large capital funding program for the University and was repeatedly asked what Purdue was doing to help American companies compete with low-cost offshore manufacturing. Hansen put the question to Dean John Hancock, who convened a special meeting of the engineering faculty. At that meeting an agreement was reached to mount a large research program that would greatly expand Barash's CIM project. This resulted in a collaboration between the EE, IE, and ME schools to form a Computer Integrated Design, Manufacturing, and Automation Center (CIDMAC), starting in 1980 with five initial $1 million grants from Cincinnati Milacron, Control Data, Cummins Engine, Ransburg, and TRW.

CIDMAC enlarged the IE effort led by Barash and Solberg with the EE expertise of Richard Paul in robotics and of King Sun Fu in artificial intelligence, along with ME's resources in computer-aided design under David Anderson. I was the executive director until 1983 when I resumed serving as head of the IE School and was succeeded by Solberg. CIDMAC's concern with the plight of American manufacturing coincided with a study by the National Academy of Engineering made under the direction of Dale Compton, a former vice president of Ford who later joined the Purdue IE School. Compton's study recommended that NSF establish and fund several national engineering research centers (ERCs) to address the problem, and the Purdue CIDMAC team, chaired by Dean Henry Yang, won one of the six ERCs awarded in 1984 for a five-year study of Intelligent Manufacturing Systems (IMS), a concept developed by Solberg and his CIDMAC associates (see Appendix 6C).

The subsequent success of the ERC/IMS project enabled Purdue to continue to get the maximum allowable NSF funding for eleven years, at which time Purdue was able to add another five years of support by winning a new, open ERC award competition. The new award was for a study of collaborative manufacturing, in which a coalition of producers and users cooperate in the design, manufacture, and distribution of new products using the most advanced networking technology. With a combined life of twenty years, the CIDMAC/ERC program grew to be the largest single research project in Purdue's history with total financial support of almost $100 million from NSF and over 65 companies. It directly employed over 1,200 students in cross-disciplinary projects and affected many thousands of other students through the creation of 60 new courses. Over 80 professors were involved in conducting research. They supervised the writing of over 200 Ph.D. theses, and published more than 1,500 technical papers. A residual effect was the refurbishing of MGL with an array of modern machine tools, robots, and FMS equipment. A notable event in MGL's history was the visit of President Ronald Reagan to Purdue in 1987 to tour the ERC facilities.

Dale Compton joined the IE faculty in 1988 as the Lillian M. Gilbreth Distinguished Professor. He taught in the areas of manufacturing, management of technology, and more recently in healthcare systems engineering. Compton served as acting head of the School in 1998-2000 and in 1999 he received the ASME Merchant Medal for Manufacturing. An excerpt from his 1997 book, *Engineering Management*, in Appendix 6D lists his recommendations for world-class manufacturing management practice.

Among Barash's Ph.D. students was Chunghorng Richard Liu, who first worked for Whirlpool Research, where he had a substantial influence on washer design before returning to Purdue in 1978. Liu worked with Barash on CIDMAC/ERC projects, and in 1984 they jointly received the ASME Blackall Award for their work on precision manufacturing. Liu's work on super hard machining, rolling contact fatigue, and nano/micro coating resulted in large savings in process costs. He made a major study of Taiwan's industry while on leave there from Purdue in 1991-94.

Other faculty who came to Purdue at the time of the CIDMAC/ ERC program included Tien-Chien Chang, Srinivasan Chandrasekar, , Chong Nam Chu, and Jay Tu. Chang was Purdue's leading researcher in the area of computer-aided process planning (CAPP). He studied feature extraction, electronic assembly, automated fixture design, tool based reasoning, and geometric tolerances with his students, as documented in his books and software that are widely used around the world. Chang's development of the "quick-turn-around" concept for the generation of prototype parts was a major contribution to the ERC program.

Chu joined the faculty in 1986 from MIT specializing in precision engineering. He left in 1992 for Seoul National University. Tu, who came from Michigan in 1992, worked in the areas of dynamic systems, laser processing, high speed spindles, precision engineering, and mechatronics, and left for North Carolina State in 2003. Chandrasekar came from Arizona State in 1986 and won an NSF Presidential Young Investigator award in 1990 and the ASME Newkirk award in 1994 for his research in materials processing, sustainable manufacturing, and nano-structured materials. He was elected a Purdue University Faculty Scholar in 1999. A more recent member of the manufacturing faculty is Gary Cheng from Columbia University.

Chandrasekar's Center for Materials Processing and Tribology continues to conduct research in an area initiated under the original CIDMAC/ERC program. Recent work with Compton on nano-structured materials and sustainable machining processes has resulted in a start-up company, M4 Sciences Corporation, which develops technologies for ultra-precision machining. As part of this project, James Mann, an IE Ph.D. student, patented a process using low-frequency vibration to achieve a many-fold productivity increase in the ultra-precise drilling of biomedical and electronic components.

Purdue's industrial engineering faculty members have been leaders in the manufacturing area and the influence of the long-lived, multi-disciplinary CIDMAC/ERC project on American and international industry has been far reaching. To commemorate his contribution, the Moshe M. Barash Distinguished Lectureship was inaugurated on March 4, 2009 with the designation of Arden L. Bement,

Director of the National Science Foundation, as its first recipient. Dr. Bement was a Purdue distinguished professor and colleague of Dr. Barash when the NSF ERC program was at its peak. As NSF's director, he manages a budget of $6 billion in support of roughly 200,000 scientists and engineers.

Modern manufacturing has come a very long way from its beginnings in the factories of the Industrial Revolution and the factory-like laboratories of Purdue's practical mechanics courses over a century ago. Purdue's manufacturing engineering efforts began at the same time that Frederick Taylor was experimenting with his new approach to factory management. Advances in computing and control theory after World War II led to the development of flexible manufacturing techniques that could make a variety of products in small batches, or an entire family of parts simultaneously on the same set of equipment, and do it in an automated, timely and cost effective way. The CIDMAC/ERC program at Purdue was a major event in the evolution of American manufacturing and a personal triumph for Professor Barash who had the foresight to imagine the shape of things to come.

Appendix 6A | William Johnson on Moshe Barash

The following is an excerpt from a paper by Dr. William Johnson, "Dr. Moshe Barash: From the Technion to Purdue: The British Connection," written for a symposium in honor of Dr. Barash held at Purdue on May 22, 1992.

I remind myself that in 1942/3, in the depths of war, when in my last year of college, at U.M.I.S.T., and before I joined the army, that our Institute had come to look more like a factory than the home of a band of scholars. One of the subjects I was then engrossed in reading, included material on *Pioneers in Scientific Management*. This was then a "non-subject," one avoided by "real, steam-boiler engineers," unfit for the current mechanical engineering curricula and in my case, something tutored only by departments thought of as scholastically suspect. How things were to change! This department was effectively industrial engineering and then lectured in commercial law, industrial relations, trade union history, and production planning and control. (There was then no possibility of such a course as this at Oxford or Cambridge. In retrospect, it proved to be a harbinger of courses for the future.)

I recall that the short book referred to above was by Urwick, a well-known practicing management consultant; he presented perhaps a dozen essays on specific management pioneers. (I write here entirely from memory.) I believe the book began with an essay on F. W. Taylor, the great innovator and pioneer of Bethlehem Steel working at the turn of the century, who left an indelible impression on me. I felt obliged after reading this essay to pay attention to his *Scientific Management*—a short book for these days and one unappreciated, I believe, and forgotten (even unknown?) by modern students of industrial engineering. I remember reading of his interest in the determination of metal cutting times (and his invention of rapid cutting steels?), culminating in the Taylor Tool Life Equation; also of tool optimizations (e.g. spade sizes) and his discussion somewhat of payment systems: maybe there were hints too of motion and time study here? (*Not* time and motion study, in that order!)

I remember the name of Henri Fayol, of Mary Parker Follet's papers (and her wise thoughts about management); of Bedeaux plans or payment systems (which precipitated a strike in the company for which my father was then working); of Gantt and his charts still in use today in highly sophisticated form through the development of the computer. And lastly, that name so dear to your hearts, Frank Gilbreth (and his wife Lillian) and his "therblig" units. . . . Finally little did I suspect that an independent meteor would pass through my life just after 1960—I refer to Moshe Barash—and that I would end my academic years spent in Gilbreth's academic home. (14-15)

APPENDIX 6B | Computerized Manufacturing

The following is an excerpt from Moshe Barash's "Computerized Manufacturing Systems for Discrete Products" in the Handbook of Industrial Engineering.

The time-sharing computer and the inexpensive minicomputer have made it possible to upgrade the concept of the numerical controlled production line into computerized manufacturing systems (CMS). A CMS is a production facility that consists of a group of process equipment units, such as machine tools or auxiliary equipment (inspection machines, washing stations, chip disposal devices, etc.), linked with an automatic materials handling system that reaches every process station, the entire facility being integrated under common computer control. Through the flexibility of NC machine tools with automatic materials handling and computer controlled production management, the CMS achieves in batch manufacturing levels of efficiency approaching those of mass production.

The latest trends in CMS design point to a greater diversity of ideas in moving the tool rather than the work piece. A system has been described that employs a large central tool magazine for 605 tools. Robots move cassettes with tools to and from individual machine magazines. . . . The overall prospect for CMS is one of accelerating growth in numbers, sophistication, and diversity of the processes included. . . . It is interesting to note that, although Japan is the first country to investigate seriously the concept of an "unmanned" factory for mechanical products, the first actual machining centers that can work for several shifts without attention were built in the United States and delivered to Sweden.

A CMS is a costly installation and careful advance planning is required in order to avoid disappointment. The problems of planning and of optimal operation of a CMS were investigated at Purdue University in a study supported by NSF. A number of design tools were developed including simulation methods for CMS and a mathematical model that requires only a limited amount of input information but that identifies bottlenecks and underutilization in a proposed CMS and provides an overall productivity estimate. The time required to perform computer analysis of one proposed system configuration using the mathematical model is on the order of a few seconds.

System scheduling and control rules were analyzed and it was found that no one rule is best for all system types. Such rules should be tested and selected through simulation. Probably the best solution is to have a simulation program included in the CMS software and delivered with it, making certain that adequate computer power is available or operating the system and for performing simulations when required. Recent developments in microprocessor technology indicate that such computing ability can be achieved at relatively low cost. (7.9.1)

APPENDIX 6C | Intelligent Manufacturing Systems

The following is an excerpt from James Solberg's "Production Planning and Scheduling in CIM."

The manufacturing environment is undergoing fundamental changes. Computer Integrated Manufacturing (CIM) and some other concepts such as Flexible Manufacturing Systems (FMS) and Just-in-Time production (JIT) are revolutionizing the way we think of structuring manufacturing organizations. We can expect to see these continue, because they are driven by competitive forces. That is, it is economically advantageous for a company to adopt any technology that works effectively in the direction of greater integration, greater flexibility, and greater responsiveness. . . . Four themes which the author believes to be important in future research in production planning and research will be mentioned. It should be apparent that these themes are consistent with one another and offer advantages in dealing with the new manufacturing realities.

The first theme is that of cooperative systems. Briefly, the cooperative systems view recognizes that very complex systems (e.g. biological, ecological, and economic systems) are beyond direct control. Instead they operate through the cooperative behavior of many interacting subsystems which may have their own independent interests, values, and modes of operation. The resulting behavior of the entire system is collectively determined. Although human systems are human creations, there may be good reason to treat them as having the kind of complexity that requires decentralized control methods.

A second related theme relates to heterarchy as an alternative to hierarchy. Although at first glance the absence of hierarchy in a complex system may be associated with chaotic behavior, deeper reflection will reveal that order can derive from well-formed, simple rules as well as it can from authority structures. . . . One method for implementing heterarchical systems is the "bulletin board" concept. A bulletin board is a software structure which provides a neutral medium for collection and exchange of data and information with all participants following a strict communication protocol. The concept has been tested in a few manufacturing cell control experiments.

The third new theme might be termed "information encapsulation." And is best expressed through object-oriented programming concepts. . . . a good object will (1) have some simple, well-defined function, (2) be usable for many different final products, and (3) present a simple, standardized interface to any present or future system. . . . These properties ensure that objects will be portable (i.e. reusable in other systems), modular (i.e. parts can be replaced with minimal side effects), and maintainable.

The fourth theme relates to real-time negotiation of resource assignments, as opposed to preplanning. This notion blends well with the three concepts mentioned above and also fits well with the dynamically varying and uncertain conditions that characterize modern manufacturing systems. Bidding, auctions, and opportunistic selection methods which have appeared in recent literature contribute to this theme.

A possible fifth theme, if it is feasible, would be some mechanism for automatic learning. The ability to improve performance, or even to keep up with environmental changes, would certainly be a desirable feature in future manufacturing software.

Putting these concepts together presents a vision of manufacturing systems that is strikingly different from that which we have seen before. Instead of machines processing jobs according to a pre-established plan and flowing under the control of a central planner, we see a population of "intelligent" entities operating in cooperation to achieve many individualized goals. For example, each machine acts to "acquire work" and each job acts to "complete required processing," while transport devices fulfill delivery objectives and so forth. The research issues that are pertinent to such a view concern the organizing principles that will insure harmonious operation and effective completion of the production work that the system exists to accomplish. (919-22)

The Spirit of the Land Grant College by Eugene Francis Savage
This mural at the entrance to Purdue University's Stewart Center shows
Education (in white robe) leading students past scenes of nineteenth-cen-
tury agriculture and technology and across a bridge to the future. In the
foreground is Abraham Lincoln signing the 1862 Land Grant Colleges Act,
introduced in Congress by Justin Smith Morrill, to Lincoln's right.

Scientists and engineers create new methods of agriculture, production,
transportation and communication. Truth, in the center, surrounded by
a coded quotation from the Roman poet Lucretius, crushes the mask of
Falsehood under her feet. At right, the benefits are being loaded onto a
ship named Abundance to be distributed to the world.
Photographs by Patrick Whalen.

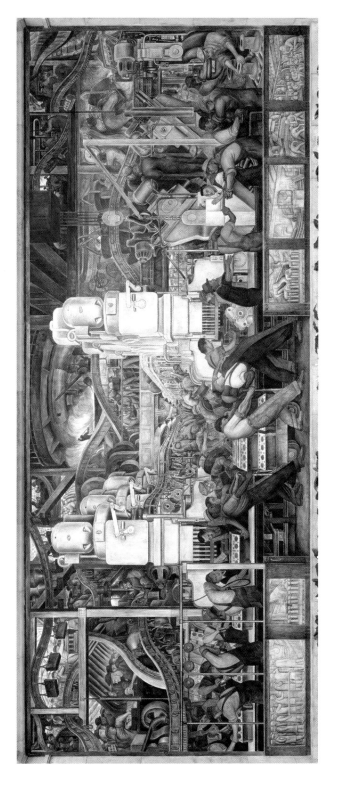

Detroit Industry: Production and Manufacture of Engine Transmission, by Diego Rivera, 1932–33, The Detroit Institute of Arts.

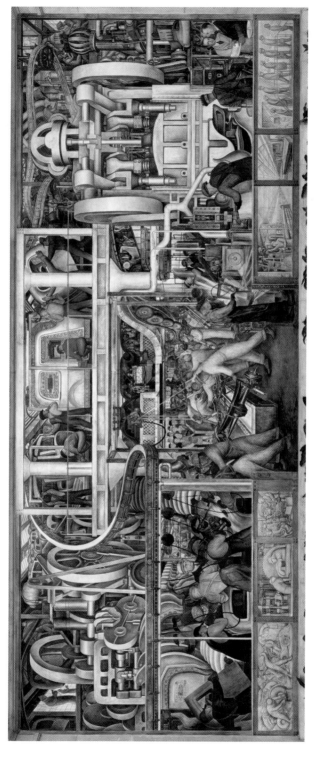

Detroit Industry: Production of Automobile Exterior and Final Assembly, by Diego Rivera, 1932-33, The Detroit Institute of Arts.

The Indiana Murals by Thomas Hart Benton
were painted for Indiana Hall at the 1933 Chicago World's Fair. The 250-
foot mural consists of twenty-two panels depicting the industrial and
cultural history of the state. Industrial Panel 10, shown above, is entitled
Electric Power, Motor-Cars, Steel.
The panel is printed courtesy of Indiana University Archives.

APPENDIX 6D | World-Class Management

The following is from Dale W. Compton's Engineering Management: Creating and Managing World Class Operations.

1. Establish as an operating goal to be world-class. Assess performance against other world-class operational functions. Use this information to establish organizational goals and objectives to be communicated to all members of the enterprise. Continuously measure and assess the performance of the system against these objectives. Regularly assess the appropriateness of the objective of attaining world-class status.

2. Recognize the importance of metrics in helping to define goals and performance expectations for the organization. Adopt or develop appropriate metrics to interpret and describe quantitatively the criteria used to measure the effectiveness of the manufacturing system and its many interrelated components.

3. Stimulate and accommodate continuous change to force experimentation and assessment of outcomes. Translate this knowledge into a framework leading to improved operational decision making. Incorporate the learning process into the fundamental operating philosophy.

4. Recognize the importance of employee involvement and empowerment as critical to achieving continuous improvement in all elements of the manufacturing system. Management's opportunity to ensure the continuity of organizational development and renewal comes primarily through the involvement of the employee.

5. Hold management responsible for a manufacturing organization's becoming world class and for creating a corporate culture committed to the customer, to employee involvement and empowerment, and to the objective of achieving continuous improvement. A personal commitment and involvement by management is critical to success.

6. Integrate all elements of the manufacturing system to satisfy the needs and wants of its customers in a timely and effective

manner. Eliminate organizational barriers to permit improved communication and to provide high quality products and services.

7. Encourage and motivate suppliers and vendors to become co-equals with other elements of the manufacturing system. This demands a commitment and expenditure of effort by all elements of the system to ensure their proper integration.

8. Instill and constantly reinforce within the organization the principle that the system and everyone in it must know their customers and must seek to satisfy the needs and wants of customers and other stakeholders.

9. Describe and understand the interdependency of the many elements of the manufacturing system. Discover new relationships. Explore the consequences of alternative decisions. Communicate unambiguously within the manufacturing organization and with its customers and suppliers. Use models as an important tool to accomplish this goal.

10. View technology as a strategic tool for achieving world-class competitiveness by all elements of the manufacturing organization. Place high priority on the discovery, development, and timely implementation of the most relevant technology and the identification and support of the people who can communicate and implement the results of research. (*Engineering Management*, 197-98)

APPENDIX 6E | Grad Courses in Manufacturing

IE 570 Manufacturing Process Engineering. Theories and applications of materials forming and removal processes in manufacturing, including product properties, process capabilities, processing equipment design, and economics. A systems approach to all aspects of manufacturing process engineering.

IE 572 Precision Manufacturing Systems. Modern manufacturing systems are critically dependent on precision machines needed to produce metal products, optical, and electronic components. This course discusses principles of precision that are critical to both existing precision manufacturing and future sub-micron/nano technology. Main topics include: computer-aided tolerancing, gaging, and process capability; principles of accuracy, repeatability, and resolution; error assessment, budgeting, and calibration; precision machine design; precision sensing and control with sub-micron resolutions; engineering economic analysis for cost-effective precision specifications.

IE 575 Computer-Aided Manufacturing I. Computer control of machines and processes in manufacturing systems. Concepts of feedback control theory as applied to NC, CNC, and DNC systems; physical elements of NC equipment for metal cutting and other processes; control hardware; NC programming languages (APT, CUTS, COMPACT, etc.), and software aspects. Process planning and adaptive control concepts as they relate to NC manufacturing. Integrated computerized systems and emerging trends. Some laboratory assignments as required.

IE 595 Projects In Machinability. Projects in metal cutting, drilling, milling, and grinding. Hours and credit to be arranged with approval of the head of the School.

IE 670 Advanced Topics in Manufacturing Engineering. Advanced research topics and approaches in manufacturing engineering, including processes and equipment. Research methodology, nontraditional processes, manufacturing systems, and competitive aspects of manufacturing.

APPENDIX 6F | A Century of Manufacturing

In 1895, U.S. factories like the one above produced 400,000 bicycles.

In 1914, Henry Ford (inset) produced 1,000 Model T Fords per day at his Highland Park factory in Detroit with 15,000 employees.

Semi-skilled workers using special tools to make Model T axles in Ford's Highland Park factory in 1914.

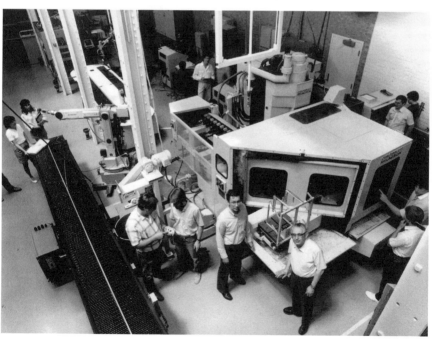

James Solberg and Moshe Barash with a computerized machining-center in the ERC CIMLAB in Michael Golden Laboratory in 1985.

Notes

1. Wiener, *Human Use of Human Beings*, 192.
2. Knoll, *Story of Purdue Engineering*, 176.
3. Knoll, *Story of Purdue Engineering*, 183.

7 | Human Factors

The practice of industrial engineering began with the manufacturing innovations of the Industrial Revolution, but it took another hundred years before it became a full-fledged profession through the combined efforts of Frederick Taylor and Frank and Lillian Gilbreth. They showed that to realize the benefits of industrialization, the introduction of new tools and processes had to be complemented by study and control of how humans interact with them as both operators and managers. Taylor and the Gilbreths launched a new field of study and professional practice aimed at understanding the manifold human factors involved and how to use that knowledge to design more beneficial industrial systems.

At first, it was thought the best way to achieve this aim was to add management courses to the business curriculum, as was done at Dartmouth, NYU, Penn, and Harvard. Unhappy with the results, Taylor then encouraged the teaching of mechanical engineering from a production systems (facilities design and process control) viewpoint as was done at Penn State, Cornell, and Purdue. This approach also disappointed because of a crucial, missing element: paying specialized attention to the man-machine interface in industrial systems. The Gilbreths were the first to champion this "human factors" approach and this became the nucleus for the new field of industrial engineering.

"People" is the first system component listed in the official definition of IE written in 1948, and twenty years later, one of the framers, Paul DeGarmo of Berkeley, said the resilience and uniqueness of the definition depended on its human component. He wrote:

> If the relationship between the human element and the system disappears then the name "industrial engineering" will indeed have to be changed, for that profession will no longer exist. But, people will be with us for many more years, I suspect, and as long as they are, industrial engineering will remain a unique and needed profession.[1]

The study of human factors in industry divided initially into two main areas, industrial ergonomics and industrial management, and the Gilbreths made pioneering contributions to both. Frank Gilbreth focused on physiology and motion study, anticipating the development of ergonomics as a multidisciplinary field with applications to a huge number of human activities. Lillian Gilbreth concentrated on the psychological and managerial aspects of human factors. The interest in industrial ergonomics took on a new importance in World War II. For example, a classic study was made by Army psychologist Alphonse Chapanis, who looked for reasons why well-trained pilots crashed functioning aircraft and found that the haphazard design of controls and their placement in the cockpit caused many pilot errors. Chapanis became a leading researcher in the field of human factors after the war when it attracted the interests of many kinds of practitioners.

The Gilbreths were first drawn to Taylor because of his fame as a management consultant. In his testimony before Congress in 1912, Taylor cited Frank Gilbreth's innovations in bricklaying as an outstanding example of scientific management. (See Appendix 7A.) Their mutual respect led to the Gilbreths installing Taylor's system at the New England Butt Co. It was there that they developed their unique approach to scientific management and industrial engineering and started to teach it in their summer school on measured functional management in 1912. The school had a major influence on the development of IE courses and curricula all over the country.

The Gilbreths' school, the first of its kind, advertised "hands-on experience with motion study by means of moving pictures and other devices."[2] In his opening lecture Frank Gilbreth discussed the four basic principles and five requirements of the Taylor system, as shown in Table 7A, taken from his teaching notes. He said this would counter the charges made by the unions that scientific management would lead to such evils as worker enslavement in robotic jobs, work speed-up and job turnover, spying on workers, worker disenfranchisement and unemployment, and destruction of the unions.

Taylor believed that time study with a stopwatch was critical to speeding up and synchronizing his micromanaged ideal factory, while Frank Gilbreth thought motion study, not time study, was the essential tool for designing optimal work systems. His use of motion pictures to make precise motion measurements led him to conclude that stopwatch methods were unreliable. Contrary to Taylor's conception that the worker must be a robotic machine minder, the

Table 7A. *An outline of Frank Gilbreth's first lecture on Taylorism in 1912*

Four Basic Principles of Scientific Management	Five Requirements of Scientific Management
1. The machine shop is a physical expression of science	1. Job study
2. Machines, materials, and workmen should be selected scientifically	2. Time standards
3. The factory should be a school of continuous instruction for workers	3. Functional management
4. Management and men should cooperate intimately in forming and executing plans	4. Standard methods
	5. Wage incentives

Gilbreths believed the greatest potential from work study would be realized when workers took responsibility for their work and used their experience, intelligence, and initiative.

In 1920, five years after Taylor's death, the Gilbreths presented a paper at the Taylor Society entitled, "An Indictment of Stop-Watch Time Study," in which they attacked Taylor's methods as being "absolutely worthless . . . misleading . . . unethical and economically wasteful."

> So far as we are concerned we learned stopwatch method of time study from Taylor, but it did not satisfy us, and consequently we had to learn from our own experience. We claim that no time study can be scientific or of any permanent value unless the motions used and the method by which the work is done are recorded and they cannot be recorded by any method that Taylor ever used. . . . [He] did not understand motion study.[3]

Lillian called for a truce after Frank Gilbreth's death in 1924 and this led to the establishment of new standards for making motion and time studies.

Work measurement was first taught at Purdue as part of the industrial design course, ME 86-87, in 1915 by Lawrence Wallace, who had attended the Gilbreth summer school in 1914. George Shepard, who took over the industrial engineering teaching duties from Wallace in 1919, was a good friend of the Gilbreths. He wrote that their way of recording and studying operations was necessary to many of the principles of management cited in his textbook.

The Gilbreths take motion pictures of the workman. Included in the picture is the dial of a clock on which the long hand moves one division every thousandth of a minute. Every film therefore records, with the workman's position, the time to one thousandth of a minute. The background of the picture is, if possible, a blackboard ruled in rectangular squares by bold white lines so that the amplitude of the motion appears against the background. If great refinement of study is required, the operator is sometimes photographed in this manner simultaneously by three cameras in three planes.[4]

As early as 1912 Shepard wrote a paper, "An Analysis of Practical Time-Motion Studies," in *Engineering Magazine,* on the steps needed to obtain both the best method and the standard time for a task. At Purdue, he introduced the courses ME 128 and 129:

ME 128 Time Studies. Method and practice in making time studies; analysis and interpretation of the data; allowances; principles of motion economy; wage payment methods; construction of formulae for rate setting.

ME 129 Micromotion Study. Preparation of standard charts from micromotion film; analysis and interpretation of data derived from micromotion film; improvement of methods; use of micromotion study as a tool of management.

Complementing Frank Gilbreth's pioneering work in ergonomics, Lillian Gilbreth was particularly interested in psychology and the study of the interplay of management and work performance to find what she and Frank called "the one best way." Due to her efforts, industrial management became the basis for a new professional school that split from engineering at Purdue in 1957. Engineering economy and production management remained with IE as core subjects, but accounting and personnel management are rarely taught in IE today. Since then, new IE research areas have emerged, like financial engineering and e-business.

Frank Gilbreth believed Taylor's idea of "functional foremanship" was very important in order for the factory to become a "school for workers." He said:

The workman must be instructed how to achieve the standard. He must have at hand a sympathetic, expert director who is teacher rather than boss. The device of functional foremanship is intended to affect this. The functional foreman teaches all of the workmen who have to perform a given function.

The foremanship of scientific management therefore requires in a given plant as many foremen as there are functions to be performed there.

The foreman of the usual organization, on the other hand, is the boss of all the men in a given room regardless of the functions performed there. He may be expert in one or more of the functions, but seldom all, and too frequently considers himself driver rather than teacher. Functional foremanship requires in a given plant as many bosses as there are departments.[5]

Taylor's functional management scheme is shown in figure 7A. It proved to be difficult to practice on a factory floor because in principle it called for every workman to have nine bosses. The more widely practiced version was the line and staff model developed by Harrington Emerson as illustrated in Figure 7B, a figure taken from George Shepard's textbook. In this version, the line foreman funneled advice to the workmen from many functional staff experts.

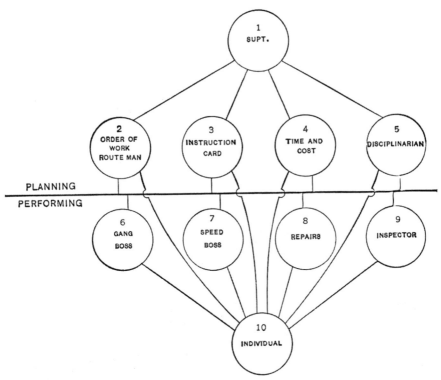

Figure 7A. *Gilbreth's graph of Taylor's functional management scheme* [6]

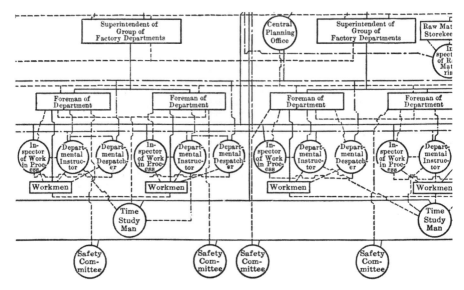

Rectangles Show Line Functions — Circles Show Staff Functions.

Figure 7B. *Detail from Shepard's diagram of a line and staff organization in a factory*

When Lillian Gilbreth came to Purdue in 1935 she joined Profs. Shepard and Hockema in teaching the popular course GE 110, and she introduced PSY 181 and the management courses GE 181-182 and GE 281-282.[7]

GE 181-182 Projects in Management. Special problems in the field of management related to industry, business, the home, and other fields of human endeavor. An honors section is open to students of superior merit who are particularly interested in management problems pertaining to industry, the home, and other fields of human endeavor.

GE 281-282 Advanced Management Problems. Specific management problems in organizations, time and motion study, industrial accounting, factory layout, economic selection of equipment, and similar topics.

PSY 181 Psychological Aspects of Industrial Efficiency. Psychological problems of industry, particularly as they relate to industrial efficiency.

Lillian Gilbreth defined her approach to management in her Ph.D. thesis, "Psychology of Management," that she wrote in 1911 and later serialized in the journal *Industrial Engineering and the En-*

gineering Digest from May 1912 to May 1913. The thesis began by defining three kinds of managers, with two being the extreme cases of a dictatorial manager on one hand and a scientific manager on the other. The third kind of manager was in between the extremes. She identified twenty important psychological implications of scientific management for the worker that are listed in Appendix 7B. She concluded that "management is fundamentally a science and must be conducted as a science, by the laboratory method, and with the most accurate of measurement, with an intensive study of the minutest details."[8]

A prominent management consultant, Ben Graham, summarized Lillian Gilbreth's view of management in the following way.

> I worked with Lillian at many conferences and recall that she often got on the subject of "top down management." Her views on the subject were very clear. She described top down improvement as enabling organizations to build beautiful, clean, simple systems that are wonderful in every way except one—they don't work. The focus of her teaching was on building healthy organizations where senior management provides the vision and the direction that guides an organization, drawing on the strength of its people where improvement comes from the bottom up.[9]

When Taylor and the Gilbreths began their work, nine out of ten people in the United States did manual work for a living—making and moving things—and it was taken for granted that the only way to produce more was to work harder. Industrial engineering showed that the smart way to obtain more production was to "work smarter," with knowledge not sweat. Today, only one out of ten U.S. workers do manual work for a living. Most work is cognitive, where people plan, design, build, and maintain semi-automatic systems that require the kind of bottom-up, self-management that the Gilbreths advocated.

Shortly after Lillian joined the faculty in 1935, the general engineering department was formed by combining the manufacturing processes and engineering drawing divisions of practical mechanics with a new industrial engineering division. Charles Beese was recruited from industry to head the large GE department with fifty staff members and nine thousand students. It soon became even larger to staff training programs for war industries. After the war Beese was replaced by Harold Bolz who went to Ohio State when general engineering was dissolved in 1955.

The original faculty team of Gilbreth, Hockema, and Shepard was augmented in 1938 with H. Barrett Rogers and Halsey Owen from industry. Robert Field, Marvin Mundel, and Leo Pigage joined in 1942 and were followed by Harold Amrine, James Greene, Oliver Hulley, Elwood Kirkpatrick, and Wallace Richardson. Rogers, Field, Mundel, Pigage, and Richardson all became prominent leaders in the field of industrial engineering at other schools, as did many of their students. The successive heads of the IE division were Shepard, Field, Mundel and Amrine.

Mundel was instrumental in starting a vigorous research program at Purdue that gained much national attention. His book, *Motion and Time Study: Improving Productivity,* was in print for over fifty years with six editions. Three of Mundel's graduate students received Purdue's Distinguished Engineering Alumni award: Robert Lehrer, Gerald Nadler, and Harold Smalley. Lehrer headed the IE program at Georgia Tech. His 1983 study, *White Collar Productivity,* identified issues at both the clerical level and the top management levels of an organization. Smalley also taught at Georgia Tech and his accomplishments in healthcare systems are discussed in chapter 8.

Nadler was successively the department head at Washington Univ. in St. Louis, the Univ. of Wisconsin, and USC. Mundel left Purdue in 1950 for a career as a consultant in the U.S. and Japan. Mundel, Lehrer and Nadler all established reputations as management consultants with popular books on industrial management, and all three were contributors to the resurgence of manufacturing in Japan after the war. In his 1983 book, *Improving Productivity and Effectiveness,* Mundel said the difference between efficiency and effectiveness was the same as the difference between doing something well and doing the right thing. He traced the "cycle-of-management" shown in Figure 7C, wherein control was established through a sequential and circular flow of nine management activities.

In his 1981 book, *The Planning and Design Approach,* Nadler defined a systems approach to project management that he said was more effective, efficient, and reliable by focusing on a realistic strategy to achieve a specific solution to a real problem. A good solution must provide for people participation, access to vital information, and anticipate future changes. In 1990, Nadler and Shozo Hibino perfected this approach in their book, *Breakthrough Thinking: Why We Must Change the Way We Solve Problems and the Seven Principles to Achieve This.* The "breakthrough" principles (listed in Appendix 5B) define a general approach to design that begins with a unique purpose and ends with a program of continuous improvement.

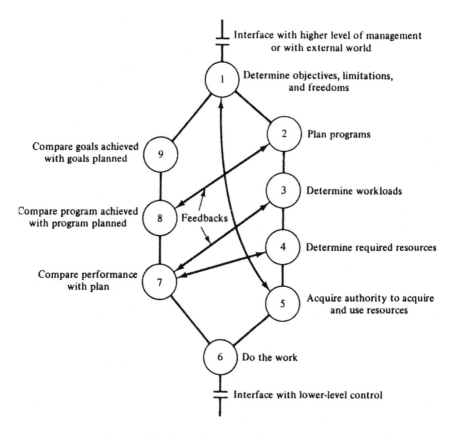

Figure 7C. *Mundel's Cycle-of-Management diagram* [10]

Harold Amrine, who succeeded Mundel as head of the IE division in 1952, taught in the areas of work measurement and production management and wrote a popular textbook, *Manufacturing Organization and Management*. Faculty in human factors and ergonomics at this time were Professors Barany, Greene, and Ritchey. James Greene came to Purdue in 1948 after graduating from Iowa. He wrote several popular textbooks on operations management and was the editor of MGraw-Hill's *Production and Inventory Control Handbook* in 1970, 1987, and 1997. John Ritchey joined the faculty in 1952 from the University of Chicago and Procter and Gamble. He taught and wrote in the area of production management and, from 1968 to 1985, worked as a management consultant with Stanford Research Institute before returning to Purdue. James Barany joined the faculty in 1958 after completing his Ph.D. under Amrine and Greene. He taught the undergraduate course in human factors and statistics and the graduate course in design of experiments for many years.

James Buck came to Purdue in 1961 from Michigan and stayed until 1981 when he left to become the IE head at the University of Iowa. He developed courses in both engineering economy and ergonomic design. Among his students were Mark Lehto and Jose Tanchoco. Tanchoco returned to Purdue in 1984 as a member of the faculty, teaching and conducting research in the area of systems engineering. Lehto went to Michigan for his Ph.D. in ergonomic design and joined the Purdue faculty in 1986. At Purdue, he specialized in decision making and safety engineering. He has conducted basic research on air bags, seat belts and helmets used for motor vehicle safety, and his book, *Warning Labels*, is a standard in the field.

In 1971, Gavriel Salvendy joined the Purdue faculty from Birmingham, England. He developed a prolific graduate program in human factors that generated fifty-one Ph.D. graduates (as documented in Appendix 10C) with support from many industrial and governmental sponsors. Notable support came from NEC of Japan that funded the NEC Chair, which Salvendy held from 1977 to 1984. Among his Ph.D. students, Richard Koubek and Vincent Duffy joined the Purdue faculty.

Salvendy was the organizer and chair of eleven international conferences on human-computer interaction between 1984 and 2005 and two on applied ergonomics in 1996 and 2008. Among his many honors, Salvendy was elected to the National Academy of Engineering in 1990, and given an honorary doctorate from the Chinese Academy of Sciences in 1995. Since 1999 he has been the head of the IE department at Tsinghua University in China and was given the Chinese Friendship Award in 2006. Two distinguished honors he received recently were the ASES John Fritz Medal in 2007 and the ASEE Imhoff Award in 2008.

Faculty members working in human factors who came in the 1980s and 90s included Antoinette Derjani, Ray Eberts, T. Govinderaj, and Richard Koubek. Derjani died tragically in an accident shortly after coming to Purdue in 1999 from Wisconsin. Eberts's interest in human-computer interaction led him to develop innovative methods of using the Internet for distance learning. Koubek came from Wright State in 1991. He returned there and later became head of IE at Penn State. Govindaraj came in 1979 from Illinois and left for Georgia Tech in 1982.

The most recent additions to the faculty included Barrett Caldwell, Vincent Duffy, Robert Feyen, Steven Landry, and Ji Soo Yi. Caldwell came in 2001 from Wisconsin and studied information flow, task design, team work, and long-duration space flights. Duffy studied at Purdue with Salvendy, taught at Hong Kong University of Science and Technology, and returned in 2005 to work in the areas of human modeling and simulation. Feyen joined the faculty in 2001, from Michigan, and studied safety engineering and computational models of human performance. Landry came in 2005 from Georgia Tech and worked in the areas of human-computer interaction and air traffic control. Yi from Georgia Tech is developing a Healthcare and Interactive Visualization Engineering Lab (HIVE) to develop and study interactive tools and techniques for improving individual and collective healthcare.

Salvendy is the editor of the popular Wiley *Handbook of Human Factors and Ergonomics* in its third edition in 2006. With 109 contributing authors and 61 chapters, the handbook provides a broad survey of the field which Salvendy divided into ten categories listed in Table 7B. The table and the complete list of chapter topics, shown in Appendix 7C, demonstrate how extensive the subject matter is and the need for a multidisciplinary approach shared by engineers, physiologists, psychologists, software engineers, equipment designers, and health professionals. The third edition required the addition of ten new chapters to keep up with developments in the field. The selected applications discussed at the end of the book include medicine, motor vehicle transportation, automation, and manufacturing and process control.

Buck and Lehto's book, *Introduction to Human Factors and Ergonomics for Engineers*, gives a more narrowly focused, textbook view of the field as it is taught in the IE program with special attention to the design of tools, tasks, and environments so that they conform to the abilities and limitations of human providers and users of products and services. The topics in Lehto and Buck's textbook are shown in Table 7C.

In his 1993 book, *Post-Capitalist Society,* Peter Drucker described the new knowledge economy that emerged from the earlier productivity revolution of Taylor and the Gilbreths. American industry is being transformed from its reliance on physical resources, financial capital, and unionized labor to being flexible, decentralized, and knowledge intensive. The shift from manual work to mental work raises fundamental problems in how to achieve the productivity increases that were characteristic of the past. Drucker said that we need an economic theory that puts knowledge into the center of the wealth-producing process that promotes continual improvement and exploitation of knowledge, as well as a quest for genuine innovation.

Table 7B. Topic categories in the *Handbook of Human Factors and Ergonomics*

The human factors function	Performance modeling
Human factors fundamentals	Evaluation
Design of tasks and jobs	Human–computer interaction
Equipment, workplace, and environmental design	Design for individual differences
	Selected applications
Design for health, safety, and comfort	

So far, no country has the educational system which the knowledge society needs. . . . Here are the new specifications: The school we need has to provide universal literacy of a high order—well beyond what "literacy" means today. It has to imbue students on all levels and of all ages with motivation to learn and with the discipline of continuing education. It has to be an open system accessible both to highly educated people and to people who . . . did not gain access to advanced education in their early years. It has to impart knowledge both as substance and as process. . . . Finally, it can no longer be a monopoly of the schools. . . . Employing organizations of all kinds—businesses, government agencies, non-profits—must become institutions of learning and teaching as well.[11]

Drucker urged the completion of the revolution in education started by engineering and medicine. Research—industrial engineering research along with others—is needed to spur the redesign of our national education system.

Table 7C. *Human factors and ergonomics topics by Lehto and Buck*

Ergonomics design	Simulation
The human system	Crews and teams
Tasks and people	Maintenance
Physical environment	Quality
Work areas and equipment	Inspection
Motion study	Safety and health
Performance measurement	Communication
Performance prediction	Control
Learning and forgetting	Decision making
Statistical sampling	Selection and training
Interviews	Compensation

The library operations research studies done at the IE School in the 1960s and 70s helped develop new technology that is emerging to cope with the efficient storage and open dissemination of vast amounts of specialized information. The Technical Assistance Program, directed by David McKinnis, provides a vital way for industry and the university to cooperate in the education process. A decade ago Ray Eberts developed Internet-based teaching programs for off-campus engineering education that extended Purdue's influence to places as far as Afghanistan, Australia, Dubai, and India.

Recently, Russell Ackoff—at the age of 88—in his 2008 book, *Turning Learning Right Side Up: Putting Education Back on Track*, urged the effort for a more revolutionary approach to education systems design. He argues, "Over the past 150 years, virtually everything has changed—except education. In the age of the Internet we educate people as we did during the Industrial Revolution."[12] Much of the Purdue IE School's current research in human-computer interaction, group decision-making, and information systems can contribute to this important movement.

Appendix 7A | Taylor on Gilbreth and Bricklaying

The following excerpt is from the Principles of Scientific Management *by Frederick Taylor.*

Bricklaying is one of the oldest of our trades. For hundreds of years there has been little or no improvement made in the implements and materials used in this trade, nor in fact in the method of laying bricks. In spite of the millions of men who have practised this trade, no great improvement has been evolved for many generations. Here, then, at least, one would expect to find but little gain possible through scientific analysis and study. Mr. Frank B. Gilbreth, a member of our Society, who had himself studied bricklaying in his youth, became interested in the principles of scientific management, and decided to apply them to the art of bricklaying. He made an intensely interesting analysis and study of each movement of the bricklayer, and one after another eliminated all unnecessary movements and substituted fast for slow motions. He experimented with every minute element which in any way affects the speed and the tiring of the bricklayer.

He developed the exact position which each of the feet of the bricklayer should occupy with relation to the wall, the mortar box, and the pile of bricks, and so made it unnecessary for him to take a step or two toward the pile of bricks and back again each time a brick is laid. He studied the best height for the mortar box and brick pile, and then designed a scaffold with a table on it, upon which all the material are placed, so as to keep the bricks, the mortar, the man, and the wall in their proper relative positions. These scaffolds are adjusted, as the wall grows in height, for all of the bricklayers by a laborer especially detailed for this purpose, and by this means the bricklayer is saved the exertion of stooping down to the level of his feet for each brick and each trowelful of mortar and then straightening up again. Think of the wasted effort that has gone on through all these years with each bricklayer lowering his body, weighing say 150 pounds, down two feet and raising it up again every time a brick (weighing about 5 pounds) is laid in the wall! And this each bricklayer did about one thousand times a day.

As a result of further study, after the bricks are unloaded from the cars, and before bringing them to the bricklayer, they are carefully sorted by the laborer and placed with their best edge up on a simple wooden frame, constructed so as to enable him to hold each brick in the quickest time and in the most advantageous position. In this way the bricklayer avoids either having to turn the brick over or end for end to examine it before laying it, and he saves also the time taken in deciding which is the best edge and end to place on the outside of the wall. In most cases, also, he saves the time taken in disentangling the brick from a disorderly pile on the scaffold. This "pack" of the bricks (as Mr. Gilbreth calls his loaded wooden frames) is placed by the helper in its proper position on the adjustable scaffold close to the mortar box.

We have all been used to seeing bricklayers tap each brick after it is placed on its bed of mortar several times with end of the handle of the trowel so as to secure the right thickness for the joint. Mr. Gilbreth found that by tempering the mortar just right, the bricks could be readily bedded to the proper depth by a downward pressure of the hand with which they are laid. He insisted that the mortar mixers should give special attention to tempering the mortar, and so save the time consumed in tapping the brick. Through all of this minute study of the motions to be made by the bricklayer in laying bricks under standard conditions, Mr. Gilbreth has reduced his movements from eighteen motions per brick to five, and even in one case to as low as two motions per brick. He has given all of the details of this analysis to the profession in the chapter headed "Motion Study," of his book entitled "Bricklaying System," published by Myron C. Clerk Publishing Company, New York and Chicago; E. F. N. Spon. of London.

An analysis of the expedients used by Mr. Gilbreth in reducing the motions of his bricklayers from eighteen to five shows that this improvement has been made in three different ways: *First*. He has entirely dispensed with certain movements which the bricklayers in the past believed were necessary, but which a careful study and trial on his part have shown to be useless. *Second*. He has introduced simple apparatus, such as the adjustable scaffold and his packets for holding bricks, by means of which, with a very small amount of cooperation from a cheap laborer, he entirely eliminates a lot of tiresome and time-consuming motions which are necessary for the bricklayer who lacks the scaffold and the packet. *Third*. He teaches the bricklayers to make simple motions with both hands at the same time, where before they completed a motion with the right hand and followed it later with one from the left hand.

For example, Mr Gilbreth teaches his bricklayer to pick up a brick in the left hand at the same instant that he takes a trowelful of mortar with the right hand. This work with two hands at the same time is, of course, made possible by substituting a deep mortar box for the old mortar board (on which the mortar spread out so thin that a step or two had to be taken to reach it) and then placing the mortar box and brick pile close together, and at the proper height on his new scaffold. These three kinds of improvements are typical of the ways in which needless motions can be entirely eliminated and quicker types of movements substituted for slow movements when scientific motion study, as Mr. Gilbreth calls his analysis, time study, as the writer has called similar work, are applied in any trade.

Most practical men would (knowing the opposition of almost all tradesmen to making any change in their methods and habits), however, be skeptical as to the possibility of actually achieving any large result from a study of this sort. Mr. Gilbreth reports that a few months ago, in a large brick building which he erected, he demonstrated on a commercial scale the great gain which is possible from practically applying his scientific study. With union bricklayers, in laying a factory wall, twelve inches thick, with two kinds of brick, faced and ruled joints on both sides of the wall, he averaged, after his selected workmen had become skillful in his new methods, 350 bricks per man *per hour*; whereas the average speed of doing this work with the old methods was, in that section of the country, 120 bricks per man per hour.

His bricklayers were taught his new method of bricklaying by their foreman. Those who failed to profit by their teaching were dropped, and each man, as he became proficient under the new method, received a substantial (not a small) increase in his wages. With a view to individualizing his workmen and stimulating each man to do his best, Mr. Gilbreth also developed an ingenious method for measuring and recording the number of bricks laid by each man, and for telling each workman at frequent intervals how many brick he had succeeded in laying.

It is only when this work is compared with the conditions which prevail under the tyranny of some of our misguided bricklayers' unions that the great waste of human effort that is going on will be realized. In one foreign city the bricklayers' union has restricted their men to *275 bricks per day* on work of this character when working for the city, and *375 per day* when working for private owners. The members of this union are probably sincere in their belief that this restriction in output is a benefit to their trade. It should be plain to all men, however, that this deliberate loafing is almost criminal, in that it results in making every workman's family pay higher rent for their housing, and also in the end drives work and trade away from their city, instead of bringing it to it.

Why is it, in a trade which has been continually practiced since before the Christian era, and with implements practically the same as they now are, that this simplification of the bricklayer's movements, this great gain, has not been made before? It is highly likely that many time during all of the years individual bricklayers have recognized the possibility of eliminating each of these unnecessary motions. But, even if in the past he did invent each one of Mr. Gilbreth's improvements, no bricklayer could increase his speed through their adoption because it will be remembered that in all cases bricklayers work together in a row and that the walls all around a building must grow at the same rate of speed. No one bricklayer, then, can work much faster than the one next to him. Nor has any one workman the authority to make other men cooperate with him to do faster work.

It is only through the *enforced* standardization of methods, *enforced* adoption of the best implements and working conditions, and *enforced* cooperation that this faster work can be assured. And the duty of enforcing the adoption of standards and of enforcing this cooperation rests with the *management* alone. The *management* must supply continually one or more teachers to show each new man the new and simpler motions, and the slower men must be constantly watched and helped until they have risen to their proper speed. All of those, who after proper teaching, either will not or cannot work in accordance with the new methods and at the higher speed must be discharged by the *management*. The *management* must also recognize the broad fact that workmen will not submit to this more rigid standardization and will not work extra hard, unless they receive extra pay for doing it. All of this involves an individual study of and treatment for each man, while in the past they have been handled in large groups.

The *management* must also see that those who prepare the bricks and the mortar and adjust the scaffold, etc. for the bricklayers, cooperate with them by doing their work just right and always on time; and they must also inform each bricklayer at frequent intervals as to the progress he is making so that he may not unintentionally fall off in his pace. Thus it will be seen that it is the assumption by the management of new duties and new kinds of work never done by employers in the past that makes this great improvement possible and that, without this new help from the management, the workmen even with full knowledge of the new methods and with the best of intentions could not attain these startling results.

Mr. Gilbreth's methods of bricklaying furnish a simple illustration of true and effective cooperation. Not the type of cooperation in which a mass of workmen on one side together cooperate with the management; but that in which several men in the management (each in his own particular way) help each workman individually, on the one hand, by studying his needs and his shortcomings and teaching him better and quicker methods, and on the other hand, by seeing that all other workmen with whom he comes in contact help and cooperate with him by doing their part of the work right and fast.

The writer has gone thus fully into Mr. Gilbreth's method in order that it may be perfectly clear that this increase in output and that this harmony could not have been attained under the management of "initiative and incentive" (that is by putting the problem up to the workman and leaving him to solve it alone) which has been the philosophy of the past. And that his success has been due to the use of the four elements which constitute the essence of scientific management.

First. The development (by the management, not the workman) of the science of bricklaying, with rigid rules for each motion of every man, and the perfection and standardization of all implements and working conditions.

Second. The careful selection and subsequent training of the bricklayers into first-class men, and the elimination of all men who refuse to or are unable to adopt the best methods.

Third. Bringing the first-class bricklayer and the science of bricklaying together, through the constant help and watchfulness of the management, and through the paying of each man a large daily bonus for working fast and doing what he is told to do.

Fourth. An almost equal division of the work and responsibility between the workman and the management. All day long the management work almost side by side with the men, helping, encouraging, and smoothing the way for them, while in the past they stood one side, gave the men but little help, and threw on to them almost the entire responsibility as to the methods, implements, speed, and harmonious cooperation.

Of these four elements, the first (the development of the science of bricklaying) is the most interesting and spectacular. Each of the three others is, however, quite as necessary for success. It must not be forgotten that back of all of this, and directing it, there must be the optimistic, determined, and hard-working leader who can wait patiently as well as work. (77-85)

APPENDIX 7B | Gilbreth on Scientific Management

In her 1914 book, The Psychology of Management, *Lillian Gilbreth set out to prove the twenty conclusions about scientific management listed below and thereby demonstrate its superiority over traditional management practices. Her study is thought to be one of the earliest examples of management theory and applied psychological research.*

1. Scientific Management is a science.
2. Scientific Management alone, of the three types of management, is a science.
3. Contrary to a widespread belief that Scientific Management kills individuality, it is built on the basic principle of recognition of the individual, not only as an economic unit but also as a personality, with all the idiosyncrasies that distinguish a person.
4. Scientific Management fosters individuality by functionalizing work.
5. Measurement in Scientific Management is of ultimate units of subdivision.
6. These measured ultimate units are combined into methods of least waste.
7. Standardization under Scientific Management applies to all elements.
8. The accurate records of Scientific Management make accurate programs possible.
9. Through the teaching of Scientific Management is unified and made self-perpetuating.
10. The method of teaching Scientific Management is a distinct and valuable contribution to education.
11. Incentives under Scientific Management not only stimulate but benefit the worker.
12. It is for the ultimate as well as the immediate benefit of the worker to work under Scientific Management.
13. Scientific Management is applicable to all fields of activity, and to mental as well as physical work.
14. Scientific Management is applicable to self-management as well as to managing others.
15. Scientific Management teaches men to co-operate with management as well as to manage.

16. Scientific Management is a device capable of use by all.

17. The psychological element of Scientific Management is the most important of all.

18. Because Scientific Management is psychologically right, it is the ultimate form of management.

19. This psychological study of Scientific Management emphasizes especially the teaching features.

20. Scientific Management simultaneously (a) increases output and wages and lowers cost, (b) eliminates waste, (c) turns unskilled labor into skilled, (d) provides a system of self-perpetuating welfare, (e) reduces the cost of living, (f) bridges the gap between the college-trained and the apprenticeship-trained worker, (g) forces capital and labor to co-operate and to promote industrial peace. (18-20)

APPENDIX 7C | Topics in Human Factors

The following list of topics is based on Gavriel Salvendy's 2006 Handbook of Human Factors and Ergonomics.

The human factors function
    The discipline of ergonomics and human factors
    Human factors engineering and systems design
Human factors fundamentals
    Sensation and perception
    Selection and control of action
    Information processing
    Communication and human factors
    Cultural ergonomics
    Decision-making models and decision support
    Mental workload and situation awareness
    Social and organizational foundations of ergonomics
    Human factors and ergonomic methods
    Anthropometry
    Basic biomechanics and workstation design
Design of tasks and jobs
    Task analysis: why, what, and how
    Task design and motivation
    Job and team design
    Personnel selection
    Design, delivery, and evaluation of training systems
    Organizational design and management
    Situation awareness
Equipment, workplace and environment design
    Affective and pleasurable design
    Workplace design
    Vibration and motion
    Sound and noise
    Illumination
Health, safety and comfort
    Occupational health and safety management
    Human error
    Ergonomics of work systems
    Psychosocial approach to occupational health

Manual materials handling
Work-related upper extremity musculoskeletal disorders
Warnings and hazard communications
Use of personal protective equipment in the workplace
Human space flight
Chemical, dust, biological, and electromagnetic radiation hazards

Performance modeling
Modeling human performance in complex systems
Mathematical models in engineering psychology
Digital human modeling for CAE applications
Virtual environments

Evaluation
Accident and incident investigation
Human factors and ergonomics audits
Cost-benefit analysis of human systems investments
Methods of evaluating outcomes

Human-computer interaction
Visual displays
Information visualization
Online communities
Human factors and information security
Usability testing
Web site design and evaluation
Design of E-business Web sites
Cognition in human-system interaction

Design for individual differences
Design for people with limitations
Design for aging
Design for children
Design for all

Selected applications
Human factors and ergonomics standards
Medicine
Motor vehicle transportation
Automation design
Manufacturing and process control
(iii-vi)

APPENDIX 7D | Grad Courses in Human Factors

IE 546 Economic Decisions in Engineering. Topics in decision making and rationality including decision analysis, decision making under uncertainty, and various descriptive and prescriptive models from operations research, economics, psychology, and business. Applications are drawn from engineering decision making, public policy, and personal decision making. Attention also is paid to designing aids to improve decision making.

IE 548 Knowledge Based Systems. Intelligent industrial systems. Expert-system and knowledge-based decision and control examples. Propositional logic, resolution principle for deduction, Horn-clause systems of logic, Dempster-Shafer uncertainty measures. Introduction to LISP and/or PROLOG. Knowledge representation schema, frames, objects and inheritance, semantic networks, rule-based representations, interface with corporate databases. Search in symbolic spaces, AND-OR trees, A-star search. Knowledge acquisition, learning by example, Kelly construct approach, neural networks. Examples of application to industrial engineering, such as manufacturing, production, etc.

IE 556 Job Design. Task analysis, personnel selection and training, job and organization design, and criteria development and use. Human factors related to job design in order to increase job satisfaction and productivity. (Also PSY 556)

IE 558 Safety Engineering. Application of human factors and engineering practice in accident prevention and the reduction of health hazards are presented. The objective of this course is to provide an understanding of the safety and health practices which fall within the responsibilities of the engineer in industry. Special attention is devoted to the detection and correction of hazards and to contemporary laws and enforcement on occupational safety and health.

IE 559 Cognitive Engineering Interactive Software. Theory and applications of software design to improve productivity and job satisfaction on information processing and cognitive tasks in the work place. Human information processing models and cognitive theories will be used to provide a theoretical basis for how to choose and display information to the user. Other topics include user-friendly displays and empirical approaches to software design. Applications of the design theory are stressed by class projects. (Also PSY 555)

IE 566 Production Management Control. Background and development of production management, plus current concepts and controls applicable to production management functions. Not open to Industrial Engineering students with a minor in management.

IE 576 Advanced Work Methods & Measurement. Advanced study of work center design and methods for improving human work. Details of work sampling, predetermined time systems, standard data development, effects of machine interference, machine pacing and fatigue on allowances, and statistical aspects of work measurement. Recent advances in work methods and measurement.

IE 577 Human Factors In Engineering. Survey of human factors in engineering with particular reference to human functions in human-machine systems, and consideration of human abilities and limitations in relation to design of equipment and work environments. (Also PSY 577)

IE 646 Advanced Decision Theory. An advanced course on the theory and models of decision making and rationality. Material includes advanced topics in decision analysis, models of probabilistic reasoning, decision making in competitive situations, and models of human and computer reasoning (including artificial intelligence).

IE 656 Research Seminar Human Factors. An in-depth study of topics of special interest in human factors. A variable title course where topics change from semester to semester. Possible topics include cognitive engineering in expert systems, human resource management, skills acquisition and retention, knowledge elicitation, and knowledge representation.

IE 659 Human Aspects of Computing. Advanced topics in ergonomic, cognitive, and social aspects of using computerized systems are discussed with regard to comfort and user satisfaction, ease and productive use of computerized systems, and effective implementation of computerized technologies in the workplace.

IE 666 Production Management Analysis. Production management policies with emphasis upon the improvement of performance in work situations. Measurement and improvement of productivity.

IE 690 Advanced Topics in Industrial Engineering. Advanced study in various fields of industrial engineering for graduate student.

Notes

1. DeGarmo, "Industrial Engineering," 70.
2. Ad for Gilbreth summer school, Gilbreth Library, School of Industrial Engineering, Purdue University.
3. Qtd. in Wrege, *Frederick W. Taylor*, 215.
4. Shepard, *Elements of Industrial Engineering*, 354.
5. Gilbreth, F., *Primer of Scientific Management*, 11.
6. Shepard,, *Elements of Industrial Engineering*, 332.
7. Purdue University, *Sixty-Fourth Annual Catalog*, 256, 320.
8. Gilbreth, *Quest of the One Best Way*, 54.
9. Graham, *People Come First*, 1.
10. Mundel, *Improving Productivity*, 24.
11. Drucker, *Post-Capitalist Society*, 197-198.
12. Ackoff, *Turning Learning Right Side Up*.

8 | Systems Engineering

The term *systems engineering* is a name that came into popular use after 1960 to describe methods of designing complex technical systems. At Bell Labs and MIT it referred to digital circuitry and software. In IE it referred to the use of operations research methods to plan large scale activities, often to provide services aimed at multiple objectives, perhaps difficult to quantify and requiring cross-disciplinary efforts. It was also used to advertise the fact that IE was no longer exclusively concerned with factory problems. Today, over a quarter of IE programs call themselves *industrial and systems engineering*, underscoring the view that systems engineering has become an integral part of the field.

Indeed, the terms *engineering* and *systems* have come to mean the same thing and the term *industrial* now is applied to both production and service systems of all kinds. The word *system* can be found in the earliest writings about industrial engineering, going back hundreds of years to Adam Smith and Charles Babbage, who described the many parts of a factory functioning in a unified way. In the Taylor era, courses in industrial engineering dealt exclusively with the design of factories as systems. The earliest well-documented production system was the Arsenal of Venice that mass-produced warships in the fifteenth and sixteenth centuries. At its peak it employed some 16,000 people to build, fit out, arm, and provision new galleys at a rate of one ship a day. It used standardized parts and production-line methods not seen again until the Industrial Revolution.

The capstone course taken by seniors in the BSIE plan of study is a systems engineering course that was first taught at Purdue in 1915 by Charles Benjamin and Lawrence Wallace as ME 23-24: *Industrial Design*. Today the same course is taught as IE 431: *Industrial Engineering Design* in very much the same way that course was originally taught. The catalog descriptions of the two courses are as follows:

184

ME 86-87 Industrial Design. The work consists in part of planning a factory layout. . . . A route chart and a model will be carefully developed, bearing in mind the important factors of internal transportation, sequence of operations, and general principles of dispatching . . . Time and motion studies will be made of actual operations and these carefully analyzed. Lectures will be given by Dean C. H. Benjamin.[1]

IE 431 Industrial Engineering Design. Capstone design experience for industrial engineering students involving analysis and synthesis of unstructured problems in practical settings. Students work in teams to formulate issues, propose solutions, and communicate results in formal written and oral presentations.[2]

A detailed description of the project work done by the students in ME 86-87 was provided by J. E. Hannum in an article entitled "Industrial Engineering at Purdue" in the *Purdue Engineering Review* of 1918. In 1915-16, Hannum wrote that four different teams worked on a new factory layout for the Sanitary Metal Basket Co., a bonus system for Rockwood Manufacturing Co., production processes for the National Fruit Juice Co. and the use of automatic screw machines at the Barbee Iron Works. In 1916-17, the class of thirteen students worked together on a study of the Monon Railroad's repair shop in Lafayette, making floor plans for every department. Hannum writes:

As far as possible all equipment was so placed so that the routing of each part would not retrace itself, but would go through from the source to the final assembly. The exact distance traveled by each part was measured . . . to show the gain in time and expense of handling. . . . The records for a period of five years were studied to determine the average length of time and cost of shipping of the different types of locomotives . . . In addition, methods of material handling, the care of equipment, the storing of all materials, . . . and provisions for the care and upkeep of tools were outlined.[3]

In 1917-18 the class of seventeen students worked together on a cost system for the Indiana Wagon Works. The study included the following steps.

(1) The collection of preliminary data; (2) making floor plans, layouts, and route charts; (3) taking of time studies, and (4) labor and material cost analysis. . . . An assembled wagon

was gone over by one of the foremen so that everyone could familiarize himself with the names and location of the parts of the wagon. The class entertained the foremen on two different evenings. . . . Floor plan layouts of every department were made and route charts showing the paths of each part were drawn. . . . Cost sheets for all the direct materials (and) for all direct labor were prepared.[4]

Table 8A. *IE 431 Industrial Engineering Design projects in 2008*

---

1. Analysis of floor-making process for modular home plant.
2. Refine capacity planning model for global auto parts manufacturer.
3. Study job fatigue factors for national retail grocery chain.
4. Justify machine tool repairs for auto parts maker.
5. Schedule job shop for sheet metal fabricator.
6. Improve workstation ergonomics for a gear producer.
7. Standardize work methods for making electrical wire harness.
8. Machining and assembly of heavy-duty brake components.
9. Design process to building lightweight trailers.
10. Redesign coating process for lighting systems fixtures.
11. Simulate handling system for lighting fixture production.
12. Design software system for reporting adverse drug events.
13. Improve efficiency for large window manufacturer.
14. Design plant layout for new steel components.
15. Develop ways to improve cancer center efficiency.
16. Improve processes for making steel doors and frames.

In comparison to the projects done in 1915-18, Table 8A lists the projects that the IE 431 students were asked to do in the spring semester of 2008. While the kinds of projects are quite similar, today's IE seniors bring a much greater and more sophisticated set of tools and techniques to bear on their work. Over the years the high professional standard expected in the design course resulted in many complaints about the amount of work required. However, alumni regularly rate IE 431 as the most important course they took as undergraduates.

Henry Gantt, a close colleague of Taylor and the Gilbreths, thought scheduling was the fundamental problem of a production system.

Two-thirds of all the difference possible between obsolete and inefficient management and the best possible one lies not in

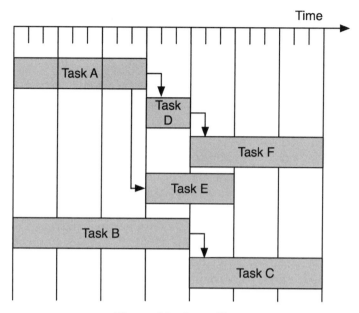

Figure 8A. *Gantt Chart*

time study work, wage-payment systems, complicated functions of control, etc., but in having all the material when you want it, where you want it, and in the condition you want it.[5]

His Gantt chart shown in Figure 8A is widely used in industry today to schedule projects.

One approach to solving Gantt's scheduling problem was the automated factory, symbolized by Henry Ford's assembly line of 1913. By 1928, transfer machines with multiple-tooling could make 10,000 automobile frames a day, with only sixteen man-minutes of direct labor needed per frame. In 1960 Ford's Cleveland engine plant automated all 530 operations that were needed to turn a rough casting into an engine block. Forty-two transfer machines were linked together to make 5,000 blocks a day with a third of the direct labor needed by a conventional system.

An alternative approach was promoted by the Japan Productivity Center in 1955 with help from consultants like Lillian Gilbreth, Mundel, Nadler, Deming and Juran. Unable to justify the costly American automated "push" system, Toyota developed a "lean, pull-not-push" system by making maximum use of worker abilities to deliver exactly what was required when and where it was needed. Parts were made in small lots with frequent and rapid changeovers in equipment requiring significant human work. High levels of efficiency, flexibility, and quality were achieved by fostering a culture of worker cooperation (quality circles) and continuous improvement ("kaizan") through worker suggestions. "Toyota out-Taylored us all," said one commentator.[6]

At Purdue, systems engineering was initially focused on the management and design of factory operations. The first industrial engineering graduate degree was awarded to John T. Elrod in 1939 for his master's thesis, *Materials Handling in Industry*, and the second Ph.D. was awarded to Nurettin Ayaasun in 1944 for a dissertation titled *Design, Location, Organization, and Industrial Management of a Munitions Plant*, supervised by Halsey Owen.

Among the first graduate students after World War II was Donald Malcolm, MSIE '48, who later developed the Program Evaluation and Review Technique (PERT) for the U.S. Navy's Polaris nuclear submarine program during the Sputnik crisis. PERT was an ingenious, modern version of the Gantt chart that became very popular for planning and scheduling complex research projects where time rather than cost was the major concern. Figure 8B illustrates the PERT concept, using the example in Figure 8A. It shows a project with five milestones (10 through 50), six activities (A through F), and two critical paths (BC and ADF) with a minimum project time of seven months. Activity E is not critical. Malcolm was honored by the Navy for his work and received a Purdue Distinguished Engineering Alumnus Award in 1965.

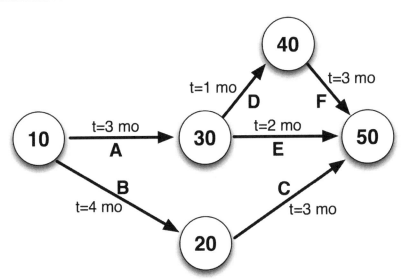

Figure 8B. *Program Evaluation and Review Technique (PERT)*

Hugh Young joined the IE faculty in 1955 and introduced the first courses in computer control and simulation applied to production systems. Other production control studies at that time were guided by Professors Paul Randolph, Thomas Bartlet, and George Brooks. Brooks and the author had worked together at DuPont on the development of the first computer system for interactive, overnight scheduling of several plants with orders arriving from different sales points across the country. Brooks later went to Georgia Tech, Young went to Arizona State, and Randolph transferred to Statistics at Purdue and then went to New Mexico State.

In the 1960s Ruddell Reed came to Purdue from Georgia Tech. His books on material handling, plant location, and layout were widely used. In 1972 Reed was awarded a chair as the Ball Brothers Professor of Engineering. He started a popular off-campus graduate program that awarded thousands of master's degrees in "industrial operations" around the state of Indiana. Colin Moodie joined the faculty in 1964 and over the next thirty-five years supervised twenty-seven Ph.D. theses in the area of production scheduling and control. He directed research projects in the EE Laboratory for Applied and Industrial Control and helped lead the large CIDMAC and ERC studies of manufacturing systems described in chapter six. In the 1970s Michael Deisenroth, Shimon Nof, and Jose Tanchoco joined Moodie to expand teaching and research in the production systems area. Deisenroth and Nof initiated research in robotics and gained much recognition for a robot emulator before Deisenroth went to Virginia Tech in 1980.

Nof focused on robotics and e-work systems, a term he coined in 1999 to describe how organizations that are globally distributed use computer networks to promote collaboration in productive work. His laboratory for Production Robotics and Integrated Software for Manufacturing and Management (PRISM) supports many interdisciplinary projects in robotics, e-work and e-business systems. PRISM is noted for its work in the automation of micro sensor networks, and for decision support in integrated production and service systems. Nof edited the *Handbook of Industrial Robotics* in 1985 and 1999, the *International Encyclopedia of Robotics* in 1988 and 1990, and the *Handbook of Automation* in 2008. He has authored books on robotics, material flow, assembly, and information technology.

Prof. Tanchoco earned his Ph.D. at Purdue under Jim Buck in 1975 and returned to Purdue to continue the work started by Moodie in manufacturing flow systems. Moodie, Nof, and Tanchoco also have been longtime advisors to Purdue's Technical Assistance Pro-

gram (TAP) that employs students to help solve production problems for small Indiana manufacturing companies. Later additions to
the systems engineering faculty were Mark Lawley, Seokcheon Lee,
Leyla Ozsen, Julie Ann Stuart, Reha Uzsoy and Yuehwern Yih. Yih
came to Purdue in 1989 and won an NSF national young investigator award in 1993 for her work on behavior-based dynamic control,
system modeling, machine learning, and artificial intelligence. She is
working with the Regenstrief Center for Healthcare Engineering to
improve patient safety, accessibility, and processing efficiency.

Uszoy joined the faculty from Florida in 1990 with research on
production control, supply chains, and semiconductor manufacturing before going to North Carolina State in 2007. Lawley came from
Illinois in 1997. His research included healthcare engineering, robust
network design, automated manufacturing, and simulation. Stuart
came to Purdue from Georgia Tech in 2000 with experience in sustainable systems engineering, and Ozsen came from Northwestern
in 2004 to work on production/logistic system design and supply-
chain management problems. Lee joined the Purdue faculty from
Penn State in 2005 with interests in computer and communication
networks, supply-chains, service systems, and ubiquitous systems.

In the 1970s there was a major shift in Purdue's systems research
effort with the arrival of Alan Pritsker to direct the Large Scale Systems (LSS) Center that was formed to develop design methodology for dealing with environmental, ecological, and societal factors.
IE professors Anderson, Olson, Petersen, Solberg, Sweet, Talavage
and their students were involved in the initial projects that included
a multidiscipline study of the John Hancock Building in Chicago.
This led to basic research on the development of computer simulation methods by Petersen, Pritsker, and Talavage. Pritsker was on the
IE faculty for twenty-five years and was a very popular teacher of
systems simulation. His company, Pritsker & Associates, employed
many students. Steven Ducket, Kenneth Musselman, Jean Reilly, and
David Wortman were among the graduates who held executive positions at P & A. Pritsker was elected to the National Academy of
Engineering in 1985, and won the IIE Gilbreth Medal in 1991, the
INFORM's Lifetime Achievement Award in 1999, and an honorary
doctorate of engineering from Purdue in 1999.

Petersen's student, Claude Pegden, developed the simulation
language, SIMAN, that became a significant competitor to SLAM.
He taught at Penn State and formed a successful consulting company
that was later acquired by Rockwell International. Pritsker's SLAM

and Pegden's SIMAN became the simulation languages most used by industrial engineers. Randall Sadowski was one of Moodie's students, who joined the faculty in 1976 and did research in production scheduling, assembly line balancing, and manufacturing automation. He left Purdue in 1988 to join Pegden's SIMAN venture.

Under Pritsker's leadership, Purdue enjoyed what his students called "the golden age of simulation" when much of the basic research was done on simulation languages like SLAM and SIMAN. Pritsker noted that simulation is particularly valuable because alternative designs can be tested without actually building a new system or disturbing a system that is already built.

> Computer simulation can be used . . . as an *explanatory device* to define a system or a problem more clearly; as an *analysis vehicle* to determine bottlenecks or other critical elements, components, and issues; as a *design accessory* to evaluate proposed solutions to design or control problems; and as a *predictor* to forecast and aid in planning future developments.[7]

> The modeling of complex production systems is often more difficult than the modeling of physical systems for the following reasons: (1) few fundamental laws are known, (2) many procedural elements are involved that are difficult to describe and represent, (3) policy inputs that are hard to quantify are required, (4) random components may be significant elements, and (5) human decision making is an integral part of such systems.[8]

The kind of projects done by Pritsker & Associate are listed in Table 8B. Appendix 8A contains a talk he gave to a sophomore class in 1993.

Another faculty member who was a pioneer in simulation development at Purdue was Stephen Roberts, who was Reed's student. Roberts wrote the simulation software INSIGHT that was used in projects for the infamous senior systems design course IE 431, which he taught for many years. IE 431 was reputed to require at least twice as much work as any other three-credit course on the Purdue campus. He left in 1990 to head the IE department at N.C. State.

Healthcare engineering, which is emerging as an important systems engineering topic in the 21st century, was studied by the Gilbreths almost a hundred years ago. In 1912 Frank did motion studies of surgical procedures and wrote a paper on the "Application of Scientific Management to the Work of the Nurse." He first wrote about

Table 8B. *Applications of computer simulation by Pritsker & Associates*

---

1. Manufacturing: plant design and layout, continuous improvement, capacity.
2. Project Planning: new products, marketing, research and development function, construction, scheduling.
3. Health Care: hospital supplies, operating room scheduling, manpower management, organ transplants.
4. Transportation: railroad performance, truck scheduling, air traffic, terminals and depots.
5. Computer/Communication Systems: performance, workflow, system reliability.
6. Financial Planning: investment, cash flow, risk assessment, balance sheet projections.
7. Environment/Ecology: flood control, pollution, energy, farm management, pest control.

---

therbligs in another paper entitled "Motion Study for the Crippled Soldier" that grew out of his experience making training films for the Army during World War I. Lillian did extensive research on people with disabilities in the 1930s and in World War II she was a leading consultant on employing women and people with disabilities in war work. She wrote a book on the subject, *Normal Lives for the Disabled*, with Edna Yost in 1945.

Lillian Gilbreth was a mentor to Purdue graduate student Harold Smalley who wrote his master's thesis, *Hospital Engineering*, in 1947, and went on to establish the field of health systems engineering. He organized the first health systems research center in the country at Georgia Tech in 1958. In 1976, he was given a Distinguished Engineering Alumnus award by Purdue. Smalley's student, John Freeman, established a health systems engineering center at Florida and hired Purdue graduate Stephen Roberts in 1968, who returned to Indiana in 1972 to direct research at the Regenstrief Institute.

In 1969 the world's largest manufacturer of dishwashers, Sam Regenstrief, established the Institute to be a research partner with the Indiana University Medical School. In his foreword to this book, Steven Beering, former dean of medicine at IU, says that Regenstrief "proposed that [IU] approach the management of health care in the same way he produced . . . dishwashers." Today the Institute is a leader in healthcare informatics with the world's largest electronic medical records system.

In 2005 the Regenstrief Foundation established the Regenstrief Center for Healthcare Engineering at Purdue with Joseph Pekny as director. The Center was inspired by the 2004 report by the National Academy of Engineering and the Institute of Medicine, *Building a Better Delivery System: A New Engineering/Health Care Partnership.* Cowritten by Dale Compton, the report outlined ways that systems engineering could help deliver safe, effective, timely, efficient, equitable, and patient-centered care. Among the contributors was IE alumnus Robert Dittus, a distinguished professor of medicine at Vanderbilt University.

IE faculty members Caldwell, Duffy, Lawley, Lehto, Muthuraman, Ozsen, Rardin, Stuart, Uzsoy, and Yih have directed over thirty student projects for the Center. Prabhu has also done work in the healthcare area using operations research to develop noninvasive methods for cancer testing and to design protein switches for nanoscale machines to retard cancer growth. Alan Pritsker's last large simulation study completed shortly before his untimely death in 2000 led to a basic change in international protocols for assigning organ transplants to waiting patients.[9] As noted above, Ji Soo Yi is currently developing a Healthcare and Interactive Visualization Engineering Lab (HIVE) to study interactive tools and techniques for improving the delivery of individual and collective healthcare.

In the spring 2008 issue of the NAE journal, *The Bridge*, Compton wrote an editorial on "Engineering and the Health Care Delivery System," in which he underscored the importance of health care as a promising area for industrial engineers. "There are abundant reasons for the problems in health care delivery," Compton wrote. "There are also abundant opportunities for working toward solving these problems. Engineers may not be able to solve all of them, but the benefits of working toward solutions can be tremendous, and the challenges they present are enormously intellectually stimulating."[10]

Finding solutions to the abundant problems of managing and designing complex systems is the fastest growing area of industrial engineering research. Purdue was an early contributor in systems research with the Large Scale Systems Center and the groundbreaking simulation work of Alan Pritsker in the 1970s and 1980s. More recently, health care has become the ideal laboratory for systems research, being comprised of interwoven networks of connected technical systems. The dire condition of the nation's health care delivery system brings added urgency to the work of industrial engineers in this area.

Appendix 8A | Pritsker on Simulation

The following is the class handout for a lecture given by Alan Pritsker at Purdue in 1993.

A model is a representation of a system which can be used as an explanatory device, an analysis tool, a design, assessor, or even a crystal ball. Simulation languages provide a framework for building models on which experiments called simulations can be performed. Through the building of models and their analysis using simulation, decision making can be supported and improved.

When exploring modeling and simulation, it is important to work with decision makers and to jointly define the purpose of the model. In this way the stage is set for managers to evaluate and implement results. In addition, it facilitates the understanding of the relative worth of system elements. Only those elements that could significantly influence decision making should be included in a model. A good rule is to start small and design the model so that it can easily be extended and embellished. Starting small may mean selecting the critical area of the system and modeling it first or it could result in building the model at an aggregate level. A simulation model should expect and encourage changes in requirements. The flexibility of the simulation approach allows for such changes.

Network models and simulation analysis tools support the above problem-solving approach. Networks are excellent vehicles for communicating system and model descriptions. They are easily modified, updated, and rebuilt. Consulting experiences demonstrate that they can be used within a team for communicating and for decomposing a large project into segments that can be assigned to team members. Networks have been used successfully to communicate model contents to government and corporate executives. Through such discussions, confidence in the model and its outputs have been increased. Discrete event logic and algebraic, difference and differential equations, can be used to augment network models.

The graphic presentation of outputs from a model is equally important (see Figures 8C and 8D). Plots, bar charts, pie charts, and histograms can be used in aiding decision making. The graphical presentation of both a model and its outputs can play a significant role in the verification and validation process. These activities lead to a greater confidence in the results of a project and, hence, support the implementation of recommendations from the project. The implementation of recommendations is an important goal for modelers, designers, and analysts.

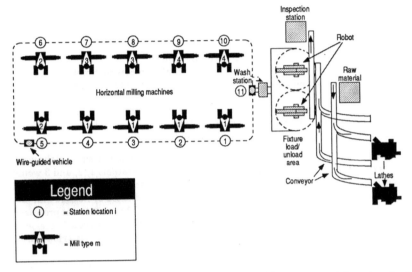

Figure 8C. *A flexible milling-machine system with raw materials going first to two lathes and then to two robots that load and unload parts to wire-guided vehicles that serve ten milling machines*

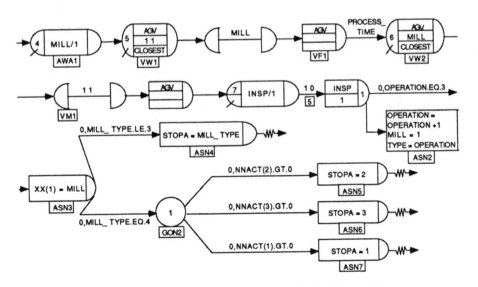

Figure 8D. *Detail of a SLAM II network model of a flexible machining system*

APPENDIX 8B | Grad Courses in Systems Engineering

IE 548 Knowledge-Based Systems. Design and control of knowledge-based systems with decision and control examples. Propositional logic, resolution principle for deduction, Horn-clause systems of logic, Dempster-Shafer uncertainty measures. Introduction to LISP and/or PROLOG. Knowledge representation schema, frames, objects and inheritance, semantic networks, rule-based representations, interface with corporate databases. Search in symbolic spaces, AND-OR trees, A-star search. Knowledge acquisition, learning by example, Kelly construct approach, neural networks. Examples of application to industrial engineering, such as manufacturing, production, etc.

IE 574 Industrial Robotics for Flexible Assembly. Design, analysis, and operation of robotic systems. System components and their control. Languages for robot control. Application design and analysis. Part feeders and tooling for robot workstations and automated assembly. Methods for planning robotic and assembly applications.

IE 579 Design and Control of Production and Manufacturing Systems. Discrete part manufacturing systems in contemporary production environments, with emphasis on flexible, demand-driven, product-based manufacturing. Currently used planning and control methodologies, such as MRP, OPT, and JIT are reviewed and integrated with appropriate facility design methodologies, including cellular design algorithms. Introduction to Computer Integrated Manufacturing (CIM) architecture and reference models and relevant control procedures, including basis approaches to appropriate data management methodologies.

IE 580 Systems Simulation. Philosophy and elements of digital simulation language. Practical application of simulation to diverse systems. Computer simulation exercises and applications are required.

IE 581 Simulation Design and Analysis. An introduction to simulation of stochastic systems on digital computers. Emphasis is on the fundamentals of simulation as a statistical experiment. Topics include uniform random numbers, input modeling, random variate generation, output analysis, variance reduction, and optimization.

IE 582 Advanced Facilities Design. Study of the theoretical and applied aspects of manufacturing systems layout. Emphasis on contemporary manufacturing, including the layout of cellular systems, automated material handling systems, and storage systems.

IE 583 Design and Evaluation of Material Handling Systems. Analysis for design and evaluation of material handling systems with emphasis on material flow control and storage. Analytic models and simulation used. Economic justification models for material handling systems.

IE 640 Network Simulation Languages. Presentation of a network approach to modeling complex systems. Design aspects of building network simulation languages such as QGERT, INS, and SAINT. Applications of network modeling and analysis.

IE 674 Computer and Communication Methods for Production Control. The study of the theoretical foundation and relevance of advanced computer and communication methods in the planning and control of intelligent production operations; manufacturing operating systems; synchronization in decentralized systems; recovery in decentralized systems; parallel processing; distributed databases; factory networks; reasoning and logic for production control.

IE 680 Advanced Simulation Design & Analysis. Continuation of IE 581 viewing stochastic simulation as a statistical experiment. Emphasis is on the research frontier. Topics include uniform random numbers, input modeling, random variate generation, output analysis, variance reduction, and optimization.

Notes

1. Purdue University, *Fortieth Annual Catalog*, 151.
2. Purdue University, *One Hundred and Thirty-Third Annual Catalog*, 200.
3. Hannum, "Industrial Engineering," 68.
4. Hannum, "Industrial Engineering," 68.
5. Gantt, *Gantt on Management*, 262.
6. Tsutsui, *Manufacturing Ideology*, 138.
7. Pritsker *Papers, Experiences, Perspectives*, 199.
8. Pritsker and Ducket, "Simulating Production Systems," 8-57.
9. Pritsker, "*Organ Transplantation*," 22.
10. Compton and Proctor, *Building a Better Delivery System*, 2.

9 | Professional Practice

Purdue industrial engineers are just about everywhere. They perform a wide variety of functions in almost every branch of American industry and every kind of public service. Most are employed as engineers but they also include teachers, plant managers, company executives, researchers, and consultants. Some are business owners, lawyers, physicians, dentists, ministers, and salespersons. There is an Army general, a harbormaster, a city manager, an FBI agent, an outfitter for mountain climbing in Nepal, a music-recording expert, a Disney World executive, and the president of a major league baseball franchise.

The number of industrial engineering degrees awarded at Purdue since 1939 are shown in Table 9A. The total number of bachelor's degrees awarded has grown from an average of 49 per year in the 1960s to 151 per year in the 2000s. Only graduate degrees were awarded before 1960 and the number of master's degrees has risen from seven per year in the 1960s to 66. While only five doctoral degrees were awarded in the first 11 years of the program, more than twice that number were earned each year by the beginning of the new century.

Table 9A. *Industrial Engineering degrees awarded at Purdue and percent women*

Years	BSIE	% Women	MSIE	% Women	Ph.D.	% Women
1939-49	0	0	73	0	5	20
1950-59	0	0	171	0	5	0
1960-69	491	0.6	232	0.4	43	0
1970-79	929	3.1	377	1.3	77	1.3
1980-89	1,701	37.4	399	20.6	101	7.9
1990-99	1,389	29.0	597	17.8	138	8.0
2000-08	1,361	27.0	596	16.9	92	15.2
Total	5,971	24.1	2,445	12.1	461	7.6

Table 9B. *IE degrees awarded in U.S. and percent women*

Date Range	BSIE	% Women	MSIE	% Women	Ph.D.	% Women
1966-69	10,428	0.5	5,506	0.6	292	1.4
1970-79	25,956	4.6	15,252	3.9	842	2.5
1980-89	39,940	26.7	16,072	15.6	994	11.2
1990-99	34,158	28.4	23,814	20.0	2,205	15.9
2000-06	**27,295**	**32.6**	**25,448**	**22.9**	**1,499**	**23.5**

*Data not available for 1999. Source: National Science Foundation 2008.

It is interesting to compare the figures for Purdue with the industrial engineering degrees granted nationwide as shown in Table 9B. The data indicate that Purdue has granted about 4% of all the BSIE degrees and 3% of all the MSIE degrees in the U.S. during the past fifty years. But for the same period, Purdue awarded 7% of the Ph.D. degrees, starting at 3% in the 1970s, rising to a high of 10% in the 1980s and then settling back to 6% of the U.S. total at the start of 21st century.

Women have earned 24%, 12%, and 8%, respectively, of the bachelor's, master's, and doctoral degrees awarded by Purdue since 1939. Women were a small minority of the engineering student body at Purdue before the 1970s when they suddenly became a very significant part of the enrollment. The percentage of women degree earners increased to current levels of 27% for bachelor's degrees, 17% for master's degrees, and 15% for Ph.D. degrees. Despite this surge in enrollment, there are still fewer female IE graduates in percentage terms at each degree level than in the nation as a whole. The Purdue figures can be compared to the national figures of 33%, 23%, and 24%, respectively. Data show that the percentages of women earning all kinds of both science and engineering degrees in the United States have been very steady at 50% for bachelor's, 44% for master's, and 38% for doctoral degrees. These numbers suggest that the number of women earning industrial engineering degrees at Purdue is likely to continue to increase at a higher rate than that of men until it approaches parity.

The plan of study for the BSIE degree is shown in Table 9C and the IE course descriptions are given in Appendix 9A. The current plan is remarkably similar to the first plan that was drawn up by School head Harold Amrine in 1955, but it is very different from the IE option that was established by Charles Benjamin and his student

Table 9C. *Plan of study for Bachelor of Science degree in Industrial Engineering*

Third Semester*	Fourth Semester	Fifth Semester
IE 230 Statistics I (3)	CE 273 Mech. Materials (3)	EE 200 Circuits (3)
MA 261 Calculus II (4)	Computing (3)	IE 335 Oper. Res. I (3)
ME 270 Mechanics (3)	IE 330 Statistics II (3)	IE 370 Mfg. Processes (3)
PHYS 241 Physics (3)	MA 262 Alg./Diff. Eqs. (4)	IE 383 Prod. Systems I (3)
General Ed. Elective (3)	General Ed. Elective (3)	General Ed. Elective (3)
Total Semester Crs.: 16	Total Semester Crs.: 16	Total Semester Crs.: 15
Sixth Semester	Seventh Semester	Eighth Semester
IE 336 Oper. Res. II (3)	IE 431 IE Design (3)	Technical Electives (12)
IE 375 Control Systems (3)	IE 443 Engr. Econ. (3)	General Ed. Elective (3)
IE 383 Prod. Systems II (3)	IE 486 Work Analysis II (3)	Total Semester Crs.: 15
IE 386 Work Analysis I (3)	Technical Electives (3)	
General Ed. Elective (3)	General Ed. Elective (3)	Total Sem. 1 & 2 Crs.: 32*
Total Semester Crs.: 15	Total Semester Crs.: 15	Total All Credits: 124

* The first and second semesters are the same for all engineering students.

Lawrence Wallace in the early 1900s, which required four courses in factory heating, lighting, sanitation, and accounting. Motivated by the ideas of Frederick Taylor, Benjamin introduced ME 23-24, Industrial Engineering in 1909 when he came to Purdue from Case Western to replace Dean Goss.

Industrial engineering students get work experience from the Cooperative Education (CO-OP) Program, Purdue's Technical Assistance Program (TAP), and summer internships in industry. CO-OP students work for companies during alternate semesters of their second and third years, and they frequently join the sponsor company after graduation. Professor Barany managed the IE School's CO-OP program for over fifty years, and he also arranged hundreds of summer internships.

Purdue's TAP program is sponsored by the State of Indiana to provide technical assistance to small businesses, and about a dozen IE faculty members over the years have monitored the work of about a hundred IE students who worked on about a thousand consulting projects around the state. It is reported that their work has resulted in millions of dollars of increased sales, reduced costs, and capital investments as well as the addition and retention of hundreds of jobs. The State of Indiana believes this benefit more than justifies the large annual contribution by the State. The author was the director of TAP for several years beginning in 1993. The current director is IE alumnus David McKinnis, Ph.D. '99.

Table 9D. *Purdue IE alumni in the United States in 2008*

Central States		Eastern States		Western States	
Indiana	1,498 (22%)	Connecticut	38	Alaska	4
Illinois	746	Washington DC	8	Arizona	117
Iowa	46	Delaware	6	California	430
Michigan	42	Florida	262	Colorado	106
Minnesota	81	Georgia	182	Hawaii	3
Ohio	422	Massachusetts	101	Idaho	11
Wisconsin	104	Maryland	78	Kansas	22
Alabama	39	Maine	7	Montana	6
Arkansas	20	N. Carolina	201	Nebraska	21
Kentucky	115	New Jersey	101	New Mexico	24
Louisiana	11	New York	139	Nevada	16
Missouri	96	Pennsylvania	156	North Dakota	6
Mississippi	11	Puerto Rico	37	Oklahoma	45
Tennessee	95	Rhode Island	6	Oregon	51
W. Virginia	8	South Carolina	64	South Dakota	3
		Virginia	158	Texas	296
		Vermont	5	Utah	17
				Washington	81
Total	**3,634 (56%)**	**Total**	**1,549 (24%)**	**Total**	**1,269 (19%)**

Total U.S. 6,452

Geographic distributions of the 7,050 IE members of the Alumni Association are shown in Tables 9D and 9E and Figures 9A and 9B. Ninety-one percent of the alumni live in the United States. Half live in the Midwest and twenty-two percent in Indiana. The remainder is divided between the East and the West, with large concentrations in California, Texas, Florida, and North Carolina. 512 alumni, nine percent of the total, live in other countries with over half of them in the Far East, notably India and Indonesia. The remaining alumni are scattered rather evenly over Central and South America, Europe, the Middle East, and Africa, giving an indication of how industrialization is spreading around the world.

Surveys show that approximately half of Purdue IEs work in production and manufacturing operations and half in business or public service. Furthermore, about half are in executive positions and half are in specialist positions. Over 75% continued their formal education with about half of those earning a master's degrees in business

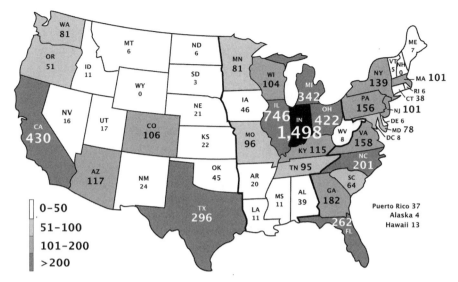

Figure 9A. *Map of Purdue IE alumni in the United States in 2008*

Table 9E. *Purdue IE alumni worldwide in 2008*

North America		Europe		South America	
Canada	15	Cyprus	4	Argentina	6
United States	6,452	France	3	Bolivia	7
Total	6,467	Germany	4	Brazil	3
Africa		Greece	1	Chile	4
Botswana	3	Ireland	2	Columbia	20
Ethiopia	1	Spain	3	Ecuador	4
Ghana	1	Switzerland	1	Paraguay	1
South Africa	1	Turkey	34	Peru	6
Tunisia	2	UK	6	Venezuela	1
Total	8	Total	58	Total	52
Central America		**Asia Minor**		**Far East**	
Bermuda	1	Bahrain	2	Australia	3
Costa Rica	9	Gaza Strip	1	China and HK	12
Dom. Republic	1	Israel	1	India	80
El Salvador	4	Jordan	9	Indonesia	77
Guadeloupe	1	Kuwait	3	Japan	3
Guatemala	9	Lebanon	4	Malaysia	26
Honduras	6	Saudi Arabia	2	Pakistan	10
Mexico	12	Syria	1	Philippines	4
Nicaragua	2	UAE	8	Singapore	23
Panama	12	Total	31	South Korea	8
Total	57			Sri Lanka	1
				Taiwan	38
				Thailand	5
				Vietnam	1
				Total	291

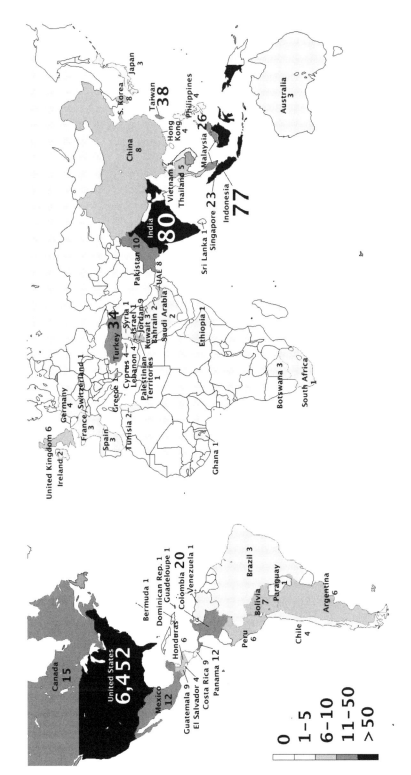

Figure 9B. *Map of Purdue IE alumni worldwide in 2008*

or management and half earning advanced degrees in engineering or a wide variety of other specialties like law and medicine. The current top employers of IE alumni are AT&T, GM, IBM, GE, Eli Lilly, Delphi, Boeing, Ford, HP, and Intel, in descending order. Included in the top 100 employers are the wide-variety of companies shown in Table 9F.

As noted earlier, Purdue alumni work in every field imaginable and some have reached the highest rungs in their organizations.

Table 9F. *Leading employers of Purdue IE alumni*

Defense/Aerospace	Frito-Lay	Microsoft
Boeing	General Mills	Texas Instruments
General Dynamics	Hillenbrand	Verizon
Goodrich	Kimberly-Clark	Xerox
Northrop-Grumman	Kraft Foods	*Conglomerates*
Halliburton	Nestle	3M
Raytheon	Pepsi	Emerson
Rolls-Royce	PPG	General Electric
US Air Force	Procter & Gamble	Ingersoll-Rand
US Army	Walt Disney	ITT
US Navy	Whirlpool	Tyco
Transport	*Business Svcs*	*Financial Svcs*
American Airlines	Accenture	AIG
FedEx	ATT	Capital One
Schneider Natl	Booz Allen	JP Morgan
UPS	Deloitte	*Capital Goods*
United Airlines	R. H. Donnelley	Caterpillar
Automotive	EDS	Deere
Allison	Ernst & Young	Trane
Chrysler	IBM	*Drugs/Healthcare*
Cummins	Oracle	Abbott Labs
Delphi	Price Waterhouse	Baxter
Ford	*Digital Technology*	Eli Lilly
General Motors	Applied Materials	Pfizer
Michelin	Cisco	Roche
Toyota	Dell	*Basic Materials*
Consumer Goods	Eaton	Alcoa
Armstrong	HP	Parker-Hannifin
Coca Cola	Intel	Intl Paper
DirecTV	Toshiba	Schlumberger
Dow Chemical	Lexmark	

There are presidents and vice-presidents, and there are school deans and department heads and persons holding distinguished teaching and research chairs. Many have been honored for their achievements by professional associations. A society to which many alumni belong is the Institute of Industrial Engineers (IIE), founded sixty years ago. Purdue faculty and graduates who have been recognized by IIE are shown in Table 9G.

Ten Purdue people have received the Institute of Industrial Engineers highest award, the Gilbreth (G) Medal for nationally recognized contributions to mankind. Eleven have served as president (P) of the Institute and forty-eight have been recognized as fellows (F)

Table 9G. *Purdue IEs honored by the Institute of Industrial Engineers*

Amrine, Harold (F)	Montreuil, Benoit (F)
Bafna, Kalash (F)	Moodie, Colin (F)
Barany, James (E,F,G,S)	Mundel, Marvin (F,G,P)
Bedworth, David (F)	Musselman,Kenneth (F,P)
Beering, Steven (H)	Nadler, Gerald (A,F,G,P)
Brookes, George (F)	Nof, Shimon (F)
Buck James (F)	Phillips, Don (A,R,F)
Chang, Tien-Chien (F,Y)	Pritsker, Alan (F,G,M,R)
Dessoukey, Maged (F)	Rardin, Ronald (A,F,P)
Ducket, Stephen (F)	Ravindran, Ravi (A,E,F)
Eskew, Michael (F,M)	Reed, Ruddell (F)
Gambrell, Charles (F,G,S)	Roberts, Stephen (F)
Gilbreth, Lillian (F,G)	Sadowski, Randall (F)
Grant, Floyd (F)	Salvendy, Gavriel (A,F)
Greene, Timothy (F,P,Y)	Schmeiser, Bruce (F,R)
Haddock, Jorge (E,Y)	Smalley, Howard (A,F,G,R)
Harmonoski, Catherine (Y)	Solberg, James (A,R)
Hoyt, Charles (F)	Sweet, Arnold (F)
Joshi, Sanjay (F,Y)	Thomas, Marlin (A,E,F,P)
Khator, Suresh (F)	Tompkins, James (A,F)
Lehrer, Robert (F,G,S)	Uzsoy, Reha (F,Y)
Leimkuhler, Ferd (F)	Webster, Dennis (F)
Malcolm, Donald (F,G,P)	Weinstein, Jeremy (P)
Malstrom, Eric (F)	Wilson, James (F,R)
Meier, Wilbur (F,P)	Wortman, David (F,P,Y)
Mize, Joe (F,G,I,P,R)	Wysk, Richard (A,F,R)
	Young, Hewitt (E,F)

Table 9H. *IE recipients of the Distinguished Engineering Alumnus Award*

Charles G. Armstrong ('03)	Robert N. Lehrer ('64)
John E. Carroll Jr. ('83)	Donald G. Malcolm ('65)
Charles T. Cotton Jr. ('85)	Joseph T. Mallof ('09)
J. Grady Cox ('71)	Joe H. Mize ('78)
J. E. de Bedout ('00)	Gerald Nadler ('75)
Donald DeFosset Jr. ('95)	Cynthia A. Niekamp ('05)
John A. Edwardson Jr. ('88)	William I. Rammes ('92)
Michael L. Eskew ('98)	John W. Reese Jr. ('80)
James L. Farlander ('86)	Rocky C. Rhodes ('97)
Robert C. Forney ('74)	Michael T. Riordan ('98)
James M. Frische ('89)	Donald A. Roach ('73)
Carroll B. Gambrell III ('69)	Robert D. Shadley ('04)
William J. Gillilan III ('03)	C. Elliot Sigal ('96)
Pedro P. Granadillo ('01)	Harold E. Smalley Sr. ('76)
Robert D. Hall ('94)	Stephen D. Spence ('79)
J. Bruce Harreld ('91)	James A. Tompkins ('99)
John R. Hodowal ('90)	Jeremy S. Weinstein ('93)
Keith J. Krach ('06)	Adel A. Zakaria ('99)

for making significant professional contributions. Other IIE awardees were honored as authors (A), educators (E), innovators (I), managers (M), researchers (R), those rendering special service (S) to the Institute, and outstanding young (Y) engineers.

The Institute has given a special honorary (H) membership award only twenty-six times in its history. In 1962 that award was given to Lillian Gilbreth and Herbert Hoover, and in 2007 former Purdue president Steven Beering was honored for his remarkable contributions to health systems engineering. In addition to having been Purdue's president and dean of medicine at Indiana University, Dr. Beering is currently chairman of the National Science Board that oversees NSF. His son John and daughter-in-law Heather are alumni of the Purdue IE School.

Each year the College of Engineering singles out a few alumni to honor with the Distinguished Engineering Alumnus (DEA) award in recognition of their career accomplishments. The thirty-five IE alumni who have been given the DEA award are listed in Table 9H and their profiles are included in Appendix 9B. Seven were recognized for their outstanding work as engineering teachers and researchers, and twenty-eight served as chairmen, presidents, CEOs, CFOs, COOs and vice presidents of mostly large enterprises making products and providing services, at the time the award was given.

The DEA awardees are chosen from a larger pool of alumni who have been identified by the different engineering school faculties as outstanding graduates of their programs. The IE faculty has recognized 86 alumni as Outstanding Industrial Engineers (OIE). They are listed in Appendix 9B. About two-thirds of them are entrepreneurs and executives of Fortune-500 companies while the remaining third are expert researchers, teachers, and consultants. Most of their organizations (57%) are in the business of making products: electronic (21%), biochemical (21%), and metal (15%). The remaining 43% provide services for clients and customers.

The highest honor given by Purdue University is the annual conferring of an honorary doctor's degree during commencement exercises. Two IE faculty members and five alumni have been given that recognition. Lillian Gilbreth was given an honorary doctorate upon her retirement from the faculty in 1948, as was Alan Pritsker in 1998 for his accomplishments as a teacher, researcher, and entrepreneur. Robert C. Forney was the first alumnus to be made an honorary doctor of engineering in 1981. He earned an MSIE in 1948 in addition to his BS and Ph.D. in chemical engineering, and went on to become executive vice president of E. I. duPont de Nemours.

James Frische, in 1994, was the second IE alumnus to receive the Dr. Eng. (Hon.). He earned a BSIE and an MSIE from Purdue before joining Sony Music where he became senior vice president of manufacturing, and founder, president and CEO of Sony's Digital Audio Disc Corporation. The third alumnus to be so honored was Michael Eskew in 2002. As chairman and CEO of United Parcel Service, he managed a company with 370,000 employees and the world's largest truck and airplane fleets, delivering some 14 million packages daily in 200 countries. The IE department endorsed giving the award to ME alumnus Donald Roach who funded a chaired professorship in manufacturing. The most recent awarding of the honorary doctorate to an IE alumnus was in 2006 when it was given to John A. Edwardson, whose distinguished career included being president and CEO of United Airlines, CFO of Northwest Airlines, and then CEO of CDW Corp., a leading provider of technology products and services for business, government, and education.

Industrial engineering alumni tend to achieve a significant measure of their professional success around the twentieth year after graduation. About a third of them become expert researchers, teachers, and consultants in engineering and other specialties. A second third are senior managers in different kinds of technical enterprises. The remainder combine executive and expert skills to be entrepreneurs, business owners, venture capitalists, and corporate leaders who deal with relatively large amounts of uncertainty when making professional decisions.

The career path of a young industrial engineer tends to touch all these bases as they follow a pattern that Barany dubbed the "four E model" of career advancement. BSIE graduates usually begin working as engineers, the first E, to learn to deal with the challenges of day-to-day operations in an industrial enterprise. Within five years, the career path generally follows one of two parallel tracks. One track hones the trainee's executive abilities by having him or her manage increasingly complex operations through line-type assignments such as foreman or supervisor. The second path develops expert skills in the design and analysis of industrial systems through assignment to staff-type jobs at both factory and corporate levels and also through further graduate study. It is common practice today to have people at this level alternate between line and staff assignments to help them find the best fit for themselves and for the organization. Such experience also lays the foundation for a specialist-generalist rising to the challenges of being an entrepreneur and taking on high-level risks and responsibilities.

In a recent alumni publication of the College of Engineering, IE alumnus Keith Krach described the function of the entrepreneur as maximizing the probability of a concept growing into a sustained enterprise and achieving "escape velocity." He said that "10 years with GM and 20 years as an entrepreneur in Silicon Valley" taught him that there were "seven key factors." He listed them as (1) finding the right market, (2) getting a very tight focus on goals, (3) gaining a clear position for action, (4) executing actions effectively, (5) using a realistic business model, (6) enlisting good people, and (7) having enough capital.[1]

Krach's factors echo Gerry Nadler's principles for *Breakthrough Thinking* shown in Appendix 5B. They also parallel Peter Drucker's recommendation that industrial organizations be re-engineered to cope with the challenges of the "knowledge revolution." The traditional baseball team or symphony orchestra approach to management needs to be replaced, Drucker said, with one that is more akin to a basketball team or jazz combo. It was Frederick Taylor's idea to have an experienced manager draw up a game plan and coordinate the efforts of a team of skilled experts in following that plan. Taylor's method was very effective for almost a hundred years. Today we need a new approach with a team of players who are both skilled in specialties and also in how to work as a team so as to adapt quickly to changing conditions, cover each other, give warnings, find openings, and initiate team action.[2]

In the 1990s the word *intrapreneur* was coined to define one who takes responsibility for turning an idea into a successful venture within an existing firm and risks being labeled a troublemaker. Steve Jobs said the Macintosh computer was an intrapreneurial venture within Apple. Joseph Pekny, Purdue's interim head of IE in 2008-09, argued that

> We need as much intrapreneurship as possible at Purdue for the faculty, students, and staff to be as empowered as possible to take measured risks. Students have to be entrepreneurial and take prudent risks to achieve their goals, or they will be left behind. Entrepreneurship is a journey, not an endpoint. It's a frame of mind.[3]

Pekny founded Advanced Process Combinatorics in 1993, a consulting company in combinatorial optimization. He heads the e-Enterprise Center in Purdue's Discovery Park.

When given awards for their accomplishments, many alumni have reflected on what it was about industrial engineering that had the biggest influence on their careers. There were five prominent themes in their testimony. First was the analytic approach to problem solving cultivated in the operations research and systems engineering courses. The second was the emphasis on the "people dimension" or the role of human leadership in marshaling the talents of others to accomplish common goals that can have huge benefits. Another theme is learning from down-to-earth, real world systems for manufacturing products and delivering services.

A fourth feature of IE that the alumni cited is the broad, interdisciplinary reach of the field across engineering and other professions and disciplines, and its involvement in international affairs. Finally, there is a sense of creative enthusiasm in having to tackle new, complex problems and be open to new ideas and developments. For example, Keith Krach said, "My days at Purdue taught me the importance of *leadership*—great things can be accomplished when you work as a team; of *constant learning*—the world of technology changes so fast you have to be constantly reeducating yourself; of *problem solving*—there is more than one solution for the big problems; and *having fun*—life is too short not to."[4]

In the second edition of the *Handbook of Industrial Engineering*, Gerry Nadler makes a detailed examination of the scope of industrial engineering. He concluded that the key ingredients that make IE unique are its human and organizational concern for the interface between technology and the people who are its managers, operators, customers, clients, and suppliers. Of great importance today is the role of the industrial engineer in working with teams of people of diverse backgrounds to achieve shared goals, increasing the likelihood that the most effective options will be adopted and implemented.

In 1989 when Nadler was president of IIE, the Institute prepared the following definition of the role of the profession in the 21st century.

> Industrial engineering will be recognized as the leading profession whose practitioners plan, design, implement, and manage integrated production and service delivery systems that assure performance, reliability, maintainability schedule adherence, and cost control. These systems will integrate people, information, material, equipment, processes, and energy throughout the life cycle of the product, service, or program.
>
> The profession will adopt as its goals profitability, effectiveness, efficiency, adaptability, responsiveness, quality, and the continuous improvement of products and services throughout their life cycles. . . . Physical, behavioral, mathematical, statistical, organizational, and ethical concepts will be used to achieve these goals.[5]

Nadler noted that while the IE profession has reached truly remarkable levels in a relatively short time since its formal beginning, there is great need for continued growth and improvement. No other profession has the breadth of IE to address such challenging problems as human-computer interaction, automation, information engineering, quality assurance, strategic planning, and environmental sustainability.

In an NAE report, *Educating the Engineer of 2020*, Purdue Dean Linda Katehi described the paradigm shift that is taking place in engineering education as the old focus on knowledge is being replaced with a new focus on skills. "Curricula based on knowledge are built from the bottom up," she wrote. "Engineers whose education is built from the bottom up cannot comprehend and address big problems.

They get lost in irrelevant details. . . . We must focus on shaping analytic skills, problem-solving skills, and design skills. We must teach methods and not solutions. We must teach future engineers to be creative and flexible, to be curious and imaginative."[6]

Appendix 9A | Undergraduate Courses in IE

IE 200 Industrial Engineering Seminar. An orientation course to inform students of the major options in the industrial engineering program.

IE 230 Probability And Statistics In Engineering I. An introduction to probability and statistics. Probability and probability distributions. Mathematical expectation. Functions of random variables. Estimation. Applications oriented to engineering problems.

IE 330 Probability And Statistics In Engineering II. Introduction to statistical inference and experimental design. Correlation, regression, single and multi-factor ANOVA, non-parametric methods. Applications to statistical quality control.

IE 332 Computing In Industrial Engineering. Introduction to computing in industrial engineering. Reinforcement of scientific programming skills, introduction to simulation and related computer tools.

IE 335 Operations Research–Optimization. Introduction to deterministic optimization modeling and algorithms. Formulation and solution of linear programs, networks flows, and integer programs.

IE 336 Operations Research–Stochastic Models. Introduction to probablistic models in operations research. Emphasis on Markov chains, Poisson processes, and their application to queueing systems.

IE 343 Engineering Economics. Cost measurement and control in engineering studies. Manufacturing cost systems. Capital investment, engineering alternatives, and equipment replacement studies.

IE 370 Manufacturing Processes I. Principal manufacturing processes; metal cutting, grinding and metal forming operations, machine tools. Nontraditional machining and welding. Computer-aided manufacturing and computer-aided graphics and design, N/C programming, robots, and flexible manufacturing systems.

IE 383 Integrated Production Systems I. Basic concepts in the design and operational control of integrated production systems. Includes topics on facility layout and material handling, material flow and information flow, resource and capacity planning, and shop floor control and scheduling.

IE 386 Work Analysis And Design I. Fundamentals of work methods and measurement. Applications of engineering, psychological, and physiological principles to the analysis and design of human work systems.

IE 431 Industrial Engineering Design. Capstone design experience for industrial engineering students involving analysis and synthesis of unstructured problems in practical settings. Students work in teams to formulate issues, propose solutions, and communicate results in formal written and oral presentations.

IE 470 Manufacturing Processes II. The interrelations of materials, processes, and design with various aspects of manufacturing.

IE 474 Industrial Control Systems. Introduction to automatic controls with reference to automation of industrial machines and processes, including linear dynamic systems, feedback control, and elements of systems analysis. Introduction to digital control.

IE 484 Integrated Production Systems II. Extensions of topics on the design and operational control of integrated production systems. Includes production databases, facility layout, material handling, advanced control and scheduling, and physical distribution. Case studies, lab assignments, and projects.

IE 486 Work Analysis And Design II. Applications of engineering, computer sciences, information sciences, and psychological principles and methods to the analysis and design of human work systems.

APPENDIX 9B | Outstanding IE Alumni

The following alumni have been honored with the Outstanding Industrial Engineer (OIE) award from the School, the Distinguished Engineering (DEA) award from the College, or the Honorary Doctor of Engineering (HDE) award from the University.

Charles G. Armstrong
BSIE '64, OIE '97, DEA '03
President and Chief Operating Officer
Seattle Mariners Baseball Club

John E. Carroll, Jr.
BSIE '64, MSIA '68, DEA '83, OIE '97
President and Chief Operating Officer
Thrall Car Manufacturing Company

Charles T. Cotton Jr.
BSIE '66, DEA '85, OIE '97
Vice President, Logistics for Operations
Frito-Lay, Inc.

J. Grady Cox
PhD '64, DEA '71, OIE '97
Professor Emeritus
Department of Industrial Engineering
Auburn University

Donald DeFosset
BSIE '71, DEA '95, OIE '97
President, Truck Group
Navistar International

John A. Edwardson, Jr.
BSIE '71, DEA '88, OIE '97, HDE '06
Chief Executive Officer
CDW Corporation

James L. Farlander
BSIE '63, MSIE '64, DEA '86, OIE '97
Vice President, Corporate Materials Mgmt
Abbott Laboratories

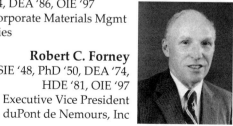

Robert C. Forney
MSIE '48, PhD '50, DEA '74,
HDE '81, OIE '97
Executive Vice President
E. I. duPont de Nemours, Inc

James M. Frische
BSIE '64, MSIE '65, DEA '89, HDE '94, OIE '97
President
Digital Audio Disc Corporation
Sony Music Entertainment, Inc.

Carroll Blake Gambrell
PhD '58, DEA '69, OIE '97
Dean Emeritus of Engineering
Mercer University

William J. Gillilan III
BSIE '68, OIE '97, DEA '03
Chairman Emeritus
Centex Corporation

Robert D. Hall
BSIE '70, DEA '94, OIE '97
President, Whirlpool Asia Appliance Grp
Executive Vice President, Whirpool Corp

J. Bruce Harreld
BSIE '72, DEA '91, OIE '97
Senior Vice President, Strategy
International Business Machines

John R. Hodowal
BSIE '66, DEA '90, OIE '97
President and Chairman of the Board
IPALCO Enterprises, Inc.

Philipp R. Hornthal
BSIE '72, MSIE '72, OIE '97
Director, Market Management
AT&T Solutions

Donald R. Juncker
BSIE '63, MSIE '65, OIE '97
Vice President of Manufacturing
Unisen, Inc.

Robert N. Lehrer
MSIE '47, PhD '49, DEA '64, OIE '97
Director Emeritus, Professor Emeritus
School of Industrial & Systems Engrg
Georgia Institute of Technology

Donald G. Malcolm
BSPSE '40, MSIE '48, DEA '65, OIE '97
Chairman, Maui Pacific Center
Vice Chairman, Maui Economic Dev Board

Joe H. Mize
MSIE '63, PhD '64, DEA '78, OIE '97
Acting Vice President for Research & Dev
Professor, Dept of Industrial Engineering
Hong Kong Univ. of Science and Tech.

Gerald Nadler
MSIE '46, PhD '50, DEA '75, OIE '97
President
The Center for Breakthrough Thinking Inc.

Dennis L. Owen
BSIE '83, OIE '97
President and CEO, Dynamic Corp
AO Quality Stamping, Inc.

William L. Rammes
BSIE '63 , DEA '92, OIE '97
Corporate Vice President,
Human Resources
Anheuser-Busch, Inc.

John W. Reese, Jr.
MSIE '54, DEA '80, OIE '97
Plant Manager, Electromagnetics Sys Div
Raytheon

Charles C. Rhodes
BSIE '75, MSIE '78, DEA '97, OIE '97
Chief Engineer
Silicon Graphics, Inc.

Donald A. Roach
BSME '52, DEA '73, HDE '95, OIE '97
Chairman
Kilburn Isotronics, Inc.

James M. Savage
BSIE '72, OIE '97
President, Physicians Health Plan

C. Elliott Sigal
BSIE '73, MSIE '73, PhD '77, DEA '96, OIE '97
President and Chief Exec Officer
Mercator Genetics

Harold E. Smalley Sr.
MSIE '47, DEA '76, OIE '97
Regent's Professor and Director
Health Systems Research Center
Georgia Institute of Technology

Stephen D. Spence
BSIE '61, DEA '79, OIE '97
President,
Measurement & Flow Control Grp
Rockwell Intl, Inc.

James A. Tompkins
BSIE '69, MSIE '70, PhD '72, OIE '97, DEA '99
President and Chairman
Tompkins Associates, Inc.

Jeremy S. Weinstein
MSIE '68, PhD '71, DEA '93, OIE '97
Vice President, Manufacturing and Tech
Whirlpool Corporation

Juan Ernesto de Bedout
BSIE '67, MSIE '68, OIE '98, DEA '00
President, Latin American Operations
Kimberly-Clark Corporation

Michael L. Eskew
BSIE '72, DEA '98, OIE '98, HDE '02
Chairman and Chief Executive
United Parcel Service

Pedro P. Granadillo
BSIE '70, OIE '98, DEA '01
Vice President, Human Resources
Eli Lilly and Company

Harry V. Huffman
BSIE '65, OIE '98
Vice President and General Counsel
Citizens Gas & Coke Utility, Indianapolis

Cynthia A. Niekamp
BSIE '81, OIE '98, DEA '05
President and General Manager
BorgWarner TorqTransfer Systems

Michael T. Riordan
BSIE '72, DEA '98, OIE '98
President and Chief Operating Officer
Fort James Corporation

John M. Stropki
BSIE '72, OIE '98
Exec Vice President &
President North America
The Lincoln Electric Company

Adel A. Zakaria
MSIE '70, PhD '73, DEA '99, OIE '98
Senior Vice President
Deere & Company

Bradford C. Anker
BSIE '68, OIE '99
Vice President, Systems Manufacturing
Acuson Corporation

Dennis Engi
BSIE '74, PhD '76, OIE '99
Chief Scientist
Sandia National Laboratories

Luis F. Machuca
BSEE '80, MSIE '81, OIE '99
Executive Vice President
Packard Bell NEC, Inc.

D. Lynn Mercer
MSIE '84, OIE '99
Product Realization Vice President
Lucent Technologies

James E. Morehouse
BSIE '66, MSIE '67, OIE '99
Vice President and Senior Account Officer
A. T. Kearney, Inc.

Curtis D. Neel, Jr.
BSIE '68, OIE '99
Senior Vice President, Logistics
Eckerd Corporation

Robert D. Shadley
BSIE '65, MSIE '66, OIE '99, DEA '04
Sr Vice President, Alliant Techsystems, Inc.
Maj General, United States Army

Robert M. Bozarth
BSIE '74, OIE '00
Vice President Facilities Sodexho
Marriott Services, Inc.

Ronald W. Dees
BSIE '72, MSIE '74, OIE '00
President
Bonne Bell, Inc.

Keith Krach
BSIE '79, OIE '00, DEA '06
Chairman and Chief Executive Officer
Ariba, Inc.

Greg Kuper
BSIE '78, OIE '00
Executive Vice President
Kimball International

Ravi Venkatesan
MSIE '87, OIE '00
Chairman
Cummins India Limited

David B. Wortman
BSIE '73, MSIE '77, OIE '00
President and Chief Executive Officer
Made2Manage Systems, Inc.

Richard Wysk
PhD '77, OIE '00
Chaired Professor
Penn State University

P. Balasubramanian
PhD '77, OIE '01
Senior Vice President
Infosys Technologies Limited

F. Hank Grant III
BSIE '73, MSIE '77, PhD '80, OIE '01
Director, Center for the Study of Wireless
Electromagnetic Compatibility
Professor, University of Oklahoma

Timothy J. Greene
BSAAE '75, MSIE '77, PhD '80, OIE '01
Dean, College of Engineering
University of Alabama

Joseph Polito
MSIE '74, PhD '77, OIE '01
Program Director, Nuclear Weapons Science
and Technology Foundations
Sandia National Laboratories

Gary W. Potts
BSIE '70, MSIE '74, OIE '01
Senior Consultant
Eli Lilly Company

Stuart C. Stock
BSIE '68, JD '71, OIE '01
Partner
Covington & Burling

Greg Schriefer
BSIE '74, MSIE '80, OIE '02
Senior Vice President for
Wire and Cable Div
Anixter, Inc.

Tamara Christen
BSIE '83, OIE '02
Senior Vice President, Marketing
Kellstrom Industries

Robert S. Dittus
BSIE '73, OIE '05
Werthan Distinguished Professor of Medicine
Chief of the Division of General
Internal Medicine, Vanderbilt Medical Center

Patricia K. Poppe
BSIE '89, MSIE '91, OIE '02
General Assembly Area Manager
General Motors Corporation

Grace Y. Lin
MS Applied Math '85, PhD '93, OIE '03
Senior Manager
T. J. Watson Research Center
International Business Machines

Robert E. Swinehart
BSIE '65, OIE '03
President and Chief Operating Officer
Emerson Power Transmission Corp.

Lee A. Chaden
BSIE '64, OIE '05
Chief Executive Officer
Branded Apparel Division (Worldwide)
Sara Lee Corporation

Thomas D. Weldon
BSIE '77, OIE '06, DEA '07
Co-founder & Chairman,
The Innovation Factory; Novoste Corp.
Co-founder & Managing Director,
Accuitive Medical Ventures

C. Dennis Pegden
BSAAE '71, MSAAE '73, PhD '75, OIE '02
Director, Development
Rockwell Software Group

Thomas J. Friel
BSIE '69, OIE '05
Chairman and Chief Executive Officer
Heidrick & Struggles International, Inc.

Joseph T. Mallof
BSIE '72, OIE '05, DEA '09
Chief Executive Officer
Ciba Vision Corporation

Mitchell M. Tseng
MSIE '73, PhD '75, OIE '05
Professor and Former Head
Dept of Industrial Eng. & Engineering Mgmt.
Hong Kong University of
Science and Technology

Patricia Anslinger
BSIE '82, OIE '07
Global Head for Mergers and Acquisitions
Accenture

Larry E. Bagwell
MSIE '69, OIE '07
Chairman, President and CEO
Rea Magnet Wire, Inc.

Jim Gibbons
BSIE '85, OIE '07
President and CEO
National Industries for the Blind

Thomas D. Kampfer
BSIE '85, OIE '07
President and Chief Operating Officer
Iomega Corporation

Daniel E. Keefe
BSIE 1981, OIE '07
Director of Asia Supply Operations
Iomega Corporation
Kodak

Kathy Kilmer
BSIE '92, OIE '07
Director, Industrial Engineering
and Measurement
Walt Disney Company

David J. Meyer
BSIE '95, OIE '07
Vice President and General Manager
Cardwell Westinghouse

William H. Pettibone
BSIE '67, OIE '07
Co-owner and Co-Chairman
Commericial Contracting Corporation

Ricardo Echevarria
BSIE '88, OIE '08
Vice President
Intel

Edward Schreck
BSIE '71, OIE '08
Chief Information Officer
Accenture

Karri L. Barbar
BSIE '82, OIE '04
Vice President of Operations
IBM Global Services Americas

W. Kee Chan
BSIE '70, OIE '05
Managing Director
Yangtzekiang Garment Manufacturing Co.
CEO, Asia Television, Hong Kong

Arturo S. Elias
BSECE '77, MSIE '78, OIE '05
President and Managing Director
General Motors Corporation de Mexico

Scott W. Givens
BSIE '88, OIE '04
President, Stadium Stunts Productions
Vice President of Entertainment, Disneyland Resort

Zone-Ching Lin
PhD '84, OIE '04
Dean of the College of Engineering
National Taiwan University of Science and
Technology

Kevin R. Scott
BSIE 1982, OIE '03
Senior Vice President, Strategy
Kraft Foods North America

Peter K. Wang
BSIE '75, OIE '05
Chairman and CEO
Tristate Holdings Limited
Hong Kong

Notes

1. Krach, "Seven Key Factors," 10.
2. Drucker, *Post-Capitalist Society*, 83.
3. Pekny, "Beyond Enterprise," 19.
4. Krach, "Seven Key Factors," 10.
5. Nadler, "The Role and Scope," 4.
6. National Academy of Engineering, *Educating the Engineer of 2020*, 154.

10 | New Directions

In a visionary study made in 2004, *The Engineer of 2020*, the National Academy of Engineering noted how the astonishing pace of technology has changed the world in the past hundred years, more than in all of preceding history and mostly for the better. More people live longer and healthier lives today, more have better means of communication and transportation and better access to education and culture, and more have better working and living conditions. This is true not only for the privileged few but for the general population of the developing world. How to bring these changes to the rest of the world is a special challenge for engineering because it directly impacts business competitiveness, military strength, healthcare, education, and the overall standard of living like that of no other profession.

The report stressed the need for engineering students to develop strong communication skills and to become actively involved in discussion over the interaction of technology and public policy. The report concluded that the attributes of the engineer of 2020 should reflect the creativity and determination of such historical figures as Albert Einstein, Martin Luther King, Pablo Picasso, Eleanor Roosevelt, and the Wright brothers. It also recommended as role models three members of the Academy: Bill Gates, for his leadership, Gordon Moore, for his problem-solving abilities, and Purdue's Lillian Gilbreth, for her ingenuity in transforming engineering by stressing its connection to human needs and capabilities. A small bio-sketch appended to the end of the report recognized her as the "mother of ergonomics" and the first woman elected to the Academy in 1966.[1]

Lillian Gilbreth was the first of nine members of the Purdue IE community who were elected to the National Academy of Engineering. They include alumni Michael Eskew, Robert Forney, Joe Mize and Gerald Nadler, and faculty members Dale Compton, Lillian Gilbreth, Alan Pritsker, Gavriel Salvendy, and James Solberg. Some of their accomplishments in advancing industrial engineering research and development have already been mentioned. A much more detailed account of IE innovation at Purdue is contained in almost five hundred dissertations conducted by doctoral students at Purdue since 1942 and listed in Appendices 10A through 10D. They are divided into the four areas described in chapters 5 through 8 above, and their works are listed in chronological order under the names of the professors who served as major advisors. Each professor and group of students defines a research cluster in a particular area and the complementary and sequential clusters in any one area trace how that field has developed. Some faculty members construct genealogical trees showing their academic descendents in subsequent generations of their students and their students' students.

Research clusters are a useful way to model how research and development ideas travel across the country and around the globe. When the people and ideas in one cluster connect with those in another cluster, they form one link in a network of links that can continue to grow in time and space. The networks enable the new ideas to travel freely and quickly throughout the worldwide research community. An example of such a network is shown in Table 10A and Figure 10A, reflecting the geographic spread of the Purdue IE research

Table 10A. *Locations of Purdue-connected researchers*

Central States	Eastern States	Western States	Other Countries
Alabama 7	Connecticut 1	Arizona 8	**Far East**
Arizona 2	Florida 4	California 4	Australia 2
Iowa 2	Georgia 8	Colorado 1	China 4
Illinois 13	Massachusetts 7	Kansas 2	India 2
Indiana —	Maryland 3	Montana 1	Korea 3
Kentucky 3	North Carolina 10	North Dakota 1	Taiwan 7
Louisiana 4	New Jersey 4	New Mexico 1	Other 2
Michigan 10	New York 6	Nevada 1	**Total 20**
Minnesota 2	Pennsylvania 9	Oklahoma 4	
Missouri 2	Rhode Island 1	Texas 3	**Other**
Ohio 8	Virginia 9	Washington 1	Canada 2
Wisconsin 3	**Total 62**	**Total 27**	Mexico 2
Total 56			Africa 3
			Europe 3
			Near East 2
			Total 12

community. This theoretical Purdue network was identified by noting the schools where Purdue IE faculty members and students did their research work, as recorded in parentheses next to their names in Appendices 10A-D. These entries were based on available records of the schools from which faculty came to Purdue and the schools to which students and faculty went. (Those coming from or going to Indiana locations were not included.)

The data in Table 10A and Figure 10A indicate how Purdue IE relates to other national and international research locations. Purdue appears to be more involved with schools in the Central and Eastern parts of the United States than it is with Western schools, although there has been a close relationship to Arizona State University that goes back over fifty years. There are strong connections with schools in the neighboring states of Illinois, Michigan, and Ohio. Other relatively important connections are with the Southeastern states of Virginia, North Carolina, Georgia, Florida, and Alabama; and the Northeastern states of Pennsylvania, Massachusetts, and New York. Outside the U.S., there has been a strong connection with the Far East, especially Taiwan (7) and China (4).

Another account of the scope and direction of modern industrial engineering research and development is the *Handbook of Industrial Engineering* edited by Gavriel Salvendy.[2] The latest edition contains 2,800 pages covering the topics that are of most interest to people practicing industrial engineering today. The topics are grouped into

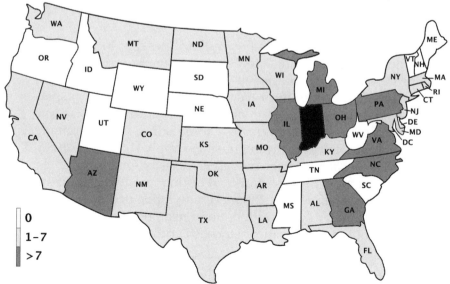

Figure 10A. *Purdue IE research connections in the United States*

three main categories: technology, people, and systems, as shown in Table 10B. The 102 chapters were written by 176 authors, of which twenty-five were Purdue faculty and alumni.

Table 10B. *Contents of the* Handbook of Industrial Engineering

---

I. Technology—(a) *Information Technology*: Building information systems, decision support, automation, computer integration technologies, networking, e-commerce, enterprise modeling; (b) *Manufacturing Technology*: Factory of the future, resource planning, automation and robotics, assembly, CAD, CAPP, CIM, JIT, clean and lean production, near-net-shape, environment, regulation, and collaboration; (c) *Service Technology*: Service systems, customers, pricing, sales, customization, client/server systems, health care, finance, retail, transport, hotels and restaurants.

II. People—(a) *Organization and Work*: Leadership, job and team design, selection and training, managing change, performance management; (b) *Human Factors and Ergonomics*: Cognitive and physical tasks, digital environments, occupational health and safety; human-computer interaction; (c) *Project Management:* Project cycle, computer-aided project management, human-centered product planning and design; design for manufacturing; managing professional services, manpower planning, methods engineering; time standards: work measurement principles and techniques.

III. Systems—*Design and Control*: Facilities size, location, layout, material-handling, storage, waste, energy, maintenance, process design, production-inventory control, scheduling, quality leadership, improvement, statistical control, inspection, reliability, service quality, standardization; (b) *Supply Chain Management*: Logistics, forecasting, transportation, warehouse networks, warehousing, supply chains; (c) *Operations Research*: Stochastic modeling; decision-making models; design of experiments; statistical testing, product cost; activity-base; cash flow; risk; inflation, simulation packages, virtual reality, linear optimization, non-linear, networks, discrete, multi-criteria optimization, stochastic processes.

An important indicator of the direction of engineering research is the program of the National Science Foundation (NSF), an inde-

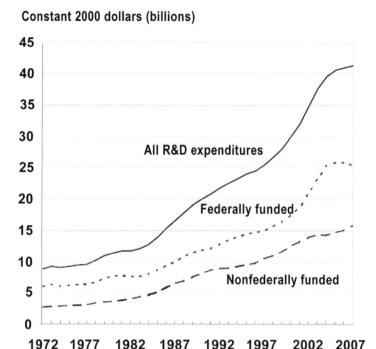

Constant 2000 dollars (billions)

45
40
35
30
25
20
15
10
5
0

All R&D expenditures

Federally funded

Nonfederally funded

1972 1977 1982 1987 1992 1997 2002 2007

Figure 10B. *U.S. academic science and engineering R&D expenditure* [4]

pendent, federal agency created by Congress in 1950 to "promote the progress of science; to advance the national health, prosperity, and welfare; to secure the national defense."[3] With an annual budget of $6 billion, NSF funds approximately 20% of the federal research at America's colleges and universities, which has grown by a factor of four in real dollars over the last thirty-five years, reaching $25 billion in 2007, of which $4.5 billion was for engineering R&D. See Figure 10B. Its $415 million R&D expenditures for science and engineering in 2007 ranked Purdue 8th nationally among universities without a medical school. (Purdue's R&D expenditures in areas other than science and engineering in 2007 was $57 million.)

The NSF Directorate of Engineering has four divisions of which one—the Civil, Mechanical, and Manufacturing Innovation (CMMI) division—focuses on advancing the disciplines of civil, industrial, manufacturing, and mechanical engineering. The CMMI division has four sets of programs, of which two are particularly pertinent to industrial engineering: Advanced Manufacturing (AM) and Systems Engineering and Design (SED). The AM programs support research in manufacturing that "emphasize efficiency, economy, and minimal environmental footprint including predictive and real-time models, novel experimental methods for manufacturing and assembly of macro-, micro-, and nano-scale devices and systems, and sensing and control techniques for manufacturing systems."[5]

Advanced Manufacturing (AM) has four programs that are concerned with manufacturing and construction machines and equipment (MCME), manufacturing enterprise systems (MES), materials processing and manufacturing (MPM), and nanomanufacturing (NM). MES covers the design, planning, and control of operations in manufacturing enterprises including analytical and computational tools for planning, monitoring, control, and scheduling of manufacturing and distribution operations. MPM comprises processing methods such as molding, forging, casting, and welding, with emphasis on environmentally benign and virtual manufacturing. NM covers nanotechnology production systems with special emphasis on its environmental, health, and societal aspects.

The Systems Engineering and Design (SED) division funds research programs in control systems (CS), dynamical systems (DS), engineering design and innovation (EDI), operations research (OR), and service enterprise systems (SES). CS supports research on control theory and control technology driven by real life applications. DS covers the analysis, modeling, simulation, and design of dynamical systems. EDI funds advances in basic design theory, tools, and software to implement design theory and new design methods that span multiple domains, such as design for the environment and for manufacturability. The OR program supports research leading to advances in the science of models and algorithms that are applicable to the operation and optimization of large-scale systems. Topic areas include optimization, simulation and stochastic modeling, and enterprise-wide models using advanced computing. SES covers strategic decision making and the modeling and analysis of service enterprises with a particular focus on healthcare and other similar public service institutions.

The Academy of Engineering study, *The Engineer of 2020*, forecasted a pattern of engineering research and development in which: (1) the pace of innovation will be rapid and accelerating; (2) technology will be more seamless, transparent, and significant; (3) the technical world will be intensely interconnected globally; and (4) success will depend heavily on social, cultural, and economic issues. With technical knowledge doubling every ten years, engineers will be challenged to create myriad new and sometimes revolutionary methods of responding to dramatic changes in the world. They will have to satisfy a variety of technical, economic, business, political, and social constraints, and frequently they must apply new technologies before the underlying science has been firmly established. The study found that the greatest challenge in engineering education is how to prepare students for this uncertainty.

In a companion report, *Educating the Engineer of 2020*, the NAE said the goal is to cultivate a core understanding of new developments in information technology, nanoscience, biotechnology, material science, and photonics, as well as in areas that are yet to be identified.[6] Advances in biotechnology have already significantly improved the quality of life, and more dramatic changes are likely. The intersection of medical knowledge and engineering has helped create pacemakers, artificial organs, prosthetic devices, imaging systems, and noninvasive surgical techniques. Developments in nanotechnology and micro-robotics are expected to repair tissue tears, clean clogged arteries, and help destroy cancers. Embedded devices will monitor organ function, and bioinformatics will customize drugs for each individual. Computer engineers are investigating virus protection architectures. Ergonomic design is an underlying theme across all engineering disciplines.

The design and manufacture of structures and materials on a molecular, or nano-scale, is being driven by the U.S. National Nanotechnology Initiative launched in 2004, with almost $1 billion in research and development funds. Nanotechnology holds great promise for detecting and correcting environmental problems. Photonics will become more significant in engineered products and systems as their physical size decreases. Biological techniques are being created that are analogous to those in microelectronics. There will be advances in fuel-cell technology, fiber optic communication, precision manufacturing, and optical sensing.

The impact of engineering on society was seen in the remarkable changes in information technology during the past few decades. The report notes how young adults today cannot imagine life without computers, mobile phones, and the Internet, while older adults, who have lived without them, may have an even deeper appreciation for them. The speed and computing power of future desktop machines is likely to make activities of present-day engineers obsolete, freeing them for more creative tasks. In some sense, everything will be "smart," as every product, every service, and every bit of infrastructure becomes more attuned to human needs.

Industrial engineers are working on the physical, psychological, and emotional interactions between information technology and humans. Computer-based design-build systems, such as was used for the Boeing 777, are becoming the norm for many production projects where multiple subsystems are combined to form a complex final product. The data and knowledge related to new technologies grow at an exponential rate, as in the health care field where more new knowledge is created in a few years than in all previous years. A special challenge is how to pursue an interdisciplinary approach to complex problems across increasingly narrow, specialized areas of knowledge.

The world's economies are interconnected, with many advanced engineering designs being created by global teams functioning across multiple time zones and cultures. New tools in manufacturing and production, new knowledge about the products being produced and their users, and the increasing ease with which information and products can be moved are changing the world. The authors of the study believed the present focus on outsourcing of low-wage mass-production jobs will shift to how to create a workforce and a business environment for new made-to-order products and services.

The combination of wireless connectivity and inventory tracking has made manufacturing and retail companies heavily dependent on new logistical methods to link together their far-flung networks of suppliers and manufacturing units. Market success or failure can hang in the balance. Outsourcing and just-in-time manufacturing have made for tightly balanced systems that allow companies to work across continents and oceans to develop products and deliver them at the right time and place around the world. The concern over natural resources is likely to be the greatest of all engineering challenges in the twenty-first century. Forty-eight countries with a total of 2.8 billion people could face freshwater shortages by 2025. The fossil fuel supply, global warming, water use, and loss of forests have all been described as crises. Engineers must develop and implement ecologically sustainable ways to achieve economic prosperity. They have to adapt solutions, in an ethical way, to the constraints of developing countries and they must consider issues of sustainability in all aspects of design and manufacturing.

In a 2007 report requested by Congress, *Rising Above the Gathering Storm: Energizing and Employing America for a Brighter Economic Future,* the Academy of Engineering described a world where the wide spread of advanced knowledge and availability of low-cost labor presented the kinds of challenges associated with the "flatness" of the modern world echoed in the books of Thomas Friedman.[7] Easy access to information technology and rising technical competences abroad have made it possible for U.S. companies to locate facilities in India, coordinate complex supply chains for manufacturing in China, and conduct "back office" service functions from almost anywhere in the world. Radiologists in India read x-rays of patients in U.S. hospitals. Architects in the United States have their drawings made in Brazil. Software is written for U.S. firms in Bangalore. Ireland has successfully implemented policies to attract research activity, as has Finland and many other countries.

The NAE report warns that, "This nation must prepare with great urgency to preserve its strategic and economic security. Because other nations have, and probably will continue to have, the competitive advantage of a low-wage structure, the United States must compete by optimizing its knowledge-based resources, particularly in science and technology, and by sustaining the most fertile environment for new and revitalized industries and the well-paying jobs they bring."[8] Furthermore, the nation must meet the pressing need for clean, affordable, reliable energy. To do this, the report recommends a course of action that depends heavily on strengthening public education by recruiting thousands of new teachers and retooling the many thousands of current teachers so as to reach millions of young minds.

The economic crisis of 2008 caused by the collapse of various financial bubbles has added a new urgency to the development challenges facing the nation and the world, but it may also provide an opportunity to make the large investments and radical infrastructure changes needed in the three key areas of energy, education, and healthcare.[9] While the need for clean, reliable, and affordable energy is necessary to sustain industrial growth, the economic benefits will not come as quickly as they may be realized in the area of healthcare, where technical innovations in managing health care information is a promising way to improve the effectiveness and efficiency of large public systems and to study how such systems can be best made to evolve. The third area of public education also promises potentially huge benefits on the order of those obtained from the Land-Grant Act of 1863 and the GI Bill of 1944, but here, too, large, long-term investments and changes in the political and technical infrastructure are needed.

The report notes that "scientists and engineers often avoid discussions" about the political obstacles to technical progress but, it says, "the stakes are too high to keep silent any longer."[10] A determining factor is the nation's commitment to compete in the global marketplace by making the U.S. workforce "the best educated, hardest working, best trained, and most productive in the world."[11] Industrial engineering can play an instrumental role in this enterprise because of its unique competence in designing and operating large-scale, human-friendly systems to produce and deliver high-technology services. The challenges are huge but the story of its past accomplishments is rich and promises a bright future for the industrial engineering profession, marking another important stage in its *enduring quest*.

Appendix 10A | Ph.D.s in Operations Research

Supervised by Prof. Remmers:

Johnson, Albert P. (1942), The Relationship of Test Scores to Scholastic Achievement for 244 Freshmen Entering Purdue University in September 1939.

Supervised by Prof. Randolph (Illinois Inst. of Tech., U. of New Mexico):

Bindschedler, Andre (1959), A Mathematical Model for Production Scheduling.

Parikh, Arvind (1960), Some Theories and Algorithms for Finding Optimal Paths Over Graphs With Engineering Applications.

Supervised by Prof. Bartlet:

White, Charles R. (1963), An Algorithm for Finding Optimal or Near Optimal Solutions to the Production Scheduling Problem. (Auburn University)

Supervised by Prof. Leimkuhler (Johns Hopkins):

Cox, Julius G. (1964), Optimum Storage of Library Material. (Auburn University)

FitzRoy, Peter T. (1966), A Quadratic Adaptive Control Model for Purposive Processes. (Monash University, Australia)

Moreno, Carlos W. (66), A Technique for Simulating Transient Sequential Queues in Production Lines.

Lister, Winston C. (1967), Least Cost Decision Rules for the Selection of Library Materials for Compact Storage.

Jain, Aridaman K. (1968), A Statistical Study of Book Use.

Sommers, Alexis N. (1968), Nondemographic Factors in V/STOL Business Travel Markets. (University of New Haven)

O'Neill, Edward T. (1970), Journal Usage Patterns and Their Implications in the Planning of Library Systems. (SUNY Buffalo)

Wu, Nesa L. (1970), Preventive Maintenance Planning and Algorithm for Scheduling Routine Maintenance Activities. (Eastern Michigan University)

Schuler, William E. (1972), A Network Model of the Systems Acquisition Process for Game Simulation.

Levy, David L. (1978), A Planning Model for the Financing of Information Centers.

Chen, Ye-Sho (1985), Statistical Models of Text Generation: A System Theory Approach. (Louisiana State University)

Supervised by Prof. Baker (Northwestern, Univ. of Cincinnati):

Hassell, H. Paul, Jr. (1968), An Analytical Design Framework for Academic Library System Formulation. (University of Alabama)

Nance, Richard E. (1968), Strategic Simulation of a Library/User/ Funder System. (Virginia Tech)

Bozoki, George (1969), Minimum-Cost Multi-Commodity Network Flows. (University of Nevada, Reno)

Yormark, Jonathan S. (1970), A Two Dimensional Resource Allocation Problem. (UCLA)

Supervised by Prof. Hill (Arizona State):

Aldrich, David W. (1969), A Decomposition Algorithm for Solving the Mixed Integer Problem. (University of North Carolina)

Kraft, Donald H. (1971), The Journal Selection Problem in a University Library System. (Louisiana State University)

Supervised by Prof. Ravindran (Berkeley, Oklahoma, Penn State):

Richmond, Tyronza R. (1971), Investigation of a Generalized Euclidean Procedure for Integer Linear.

Arthur, Jeffrey L. (1977), Contributions to the Theory and Applications of Goal Programming.

Balasubramanian, Parasuram (1977), Optimal Resource Utilization in Communicable Disease Control With Special Reference to Syphilis Control.

Philipson, Roland H. (1979), An Algorithm for the Optimization of a Multiple-Spindle Automatic Bar Machine.

Sadagopan, Sowmyanarayanan (1979), Multiple Criteria Mathematical Programming–A Unified Interactive Approach. (Indian Institute Of Technology Bangalore)

Gupta, Omprakash (1980), Branch and Bound Experiments in Nonlinear Integer Programming.

Hafez, Enayat I. (1980), Mathematical Programming Applied to Power Systems Expansion Planning with Pumped Storage.

Eswaran, P.K. (1983), Interactive Decision Making with Multiple Criteria-Algorithms and Applications.

Supervised by Prof. Solberg (Michigan, Toledo):

Kelly, John C. (1974), The Theory of Repetition Networks With Application to Computer Programs.

Engi, Dennis (1976), A Graph-Theoretic Decomposition of Markovian Networks. (Purdue University)

Jones, Albert T. (1978), Queueing Theory in Air Transportation Systems.

Grant, Floyd H. (1980), Variance Reduction Techniques in Stochastic Shortest Route Analysis. (University of Oklahoma)

Stecke, Kathryn E. (1981), Production Planning Problems for Flexible Manufacturing Systems. (University of Michigan)

Fox, Dale R. (1983), Parametric and Algorithmic Solutions for the Steady-State Analysis of Markovian Queueing Systems using Graphical Enumeration.

Vinod, Balakrishnan (1983), Queueing Models for Flexible Manufacturing Systems Subject to Resource Failure.

Maley, James (1987), A Combinatorial Optimization Solution Strategy with Application to Transport Systems.

Heim, Joe (1990), Integration of Distributed Heterogeneous Simulation Models for Design of Manufacturing Systems.

Lin, Grace Yuh-Jiun (1993), A Distributed Production Control for Intelligent Manufacturing Systems.

Moses, Scott A. (1995), Adaptive Closed-Loop Production Control for Discrete Manufacturing Systems.

Supervised by Prof. Morin (Northwestern):

Polito, Joseph (1977), Distribution Systems Planning in a Price Responsive Environment.

Akileswaran, Vaidyanathan (1979), Optimality Analysis of Heuristic Decision Rules In Water Resource Planning.

Evans, Gerald W. (1979), Optimal Generation Planning for Electric Utilities. (University of Louisville)

Hiatt, William H. (1979), A Group Theoretic Approach to Parametric Integer Programming.

Preklas, David M. (1979), A Unifying Theory of Algorithms for Discrete Optimization Problems.

Hazen, Gordon B. (1980), A Convexity-Based Efficient-Point Strategy for Multicriteria Optimization. (Northwestern University)

Klein, Cerry M. (1983), Duality in Dynamic Programming. (University of Missouri)

Nagaraj, Kolinjuwadi (1983), The A Posteriori Approach for Discrete Optimization.

Kincaid, Rex K. (1984), The Location of Central Structures in Graphs. (College of William & Mary)

Trafalis, Theodore (1989), Efficient Faces of a Polytope: Interior Methods in Multi-Objective Optimization.

Kaiser, Mark (1991), Centers of Convex Bodies. (Louisiana State University)

Zhang, Zaili (1995), Topics in Linear, Dynamic and Multi-Objective Optimization.

Giffen, Bill (2004), Deterministic & Stochastic Extensions of the Jeep Problem.

Supervised by Prof. Sweet (Purdue):

Young, Robert E. (1977), Transition Probability Estimation for Discrete State Markov Chains. (North Carolina State University)

Flanigan, Mary A. (1993), A Flexible, Interactive, Graphical Approach to Modeling Stochastic Input Processes.

Supervised by Prof. Sparrow (Michigan, Johns Hopkins, Univ. of Houston):

Haddock-Acevedo, Jorge (1981), Energy Planning for Puerto Rico; A Systems Modeling Approach. (University of Richmond)

Kachitvichyanukak, (1982), Computer Generation of Poison, Binomial and Hypergeometric Random Variates.

Kilmartin, Michael (1982), The Value of Electricity Freed by Distributed Solar Systems.

Malakooti, Behnam (1982), An Interactive Paired Comparison Method for Multiple Criteria Decision Making Optimization. (Case Western Reserve University)

Cochran, Jeffery K. (1984), Improving the Productivity of the American Automobile: A Stochastic Process View. (Air Force Institute of Technology)

Huang, Shiu-Ling (1987), Expert Systems for Grading Hardwood Lumber.

Kem, Dale (1988), Optimal Distribution Network Configuration for the Single-Commodity, Single-Supplier, W-Distribution Point, N-Retailer System.

Parker, Stephen (1994), Military Force Structure and Realignment through Dynamic Simulation.

Lee, Julien (1995), Integration of the Steel and Electricity Industries Using Price and Load Information Exchange.

Boerger, Peter (1996), Engineering Economic Model of Optimal Manufacturing Technology Selection in a Framework of Concurrent Engineering: Theory and Application to Industrial Electrification.

Brady, Tom (1996), Prescriptive Simulation: A Heuristic Approach. (Purdue University)

Leung, Thomas (1997), The World Iron and Steel Industry and Its Impacts on Indiana Iron and Steel and Electric Utility Industries.

Alkhal, Farqad (1998), Artificial Neural Networks for Long-Term Electric Load Modeling and Forecasting.

Bowen, Brian (1998), Short-Term Benefits from Central Unit Commitment and Dispatch: Application to the Southern African Power Pool.

Hutchison, Steve (1998), The War on Drugs: Mathematical Modeling for Policy Selection and Resource Allocation.

Stamber, Kevin (1998), Forecasting Industrial Motor Stock: Economic Theory and Policy Analysis.

Wang, Jenhung (1998), The Value of Perfect Information in Power System Planning Problems.

Lin, Pei-chun (1999), Estimating the Energy Demand of Industrial HVAC.

Nderitu, David G. (1999), Electric Utility Capacity Expansion Planning Model Incorporating Power Flow Equations.

Al-Salamah, Muhammad (2004), A Cross-Border Natural Gas Supply System to Support the Electricity Industry in the Countries of the Gulf Cooperation Council (GCC): A Large Scale Mixed Integer Linear Mathematical Programming Model of the Production, Transportation, and Storage of Natural Gas.

Naidoo, Ramu (2004), Planning Models for Electric Transmission Network Expansion.

Siriariyaporn, Veeradech (2005), Reducing the Cost of Serving Uncertain Loads via the Portfolio of Supply Contracts.

Supervised by Prof. Schmeiser (Georgia Tech):

Dattero, Ronald S. (1982), Stochastic Models from Event Count Data. (Florida Atlantic University)

Swain, James J. (1982), Monte Carlo Estimation of the Sampling Distribution of Non-Linear Parameter Estimators. (University of Alabama)

Nelson, Barry (1983), Variance Reduction in Simulation Experiments: A Mathematical-Statistical Framework. (Northwestern University)

Kang, Keebom (1984), Confidence Interval Estimation via Batch Means and Time Series Modeling. (U.S. Naval Postgraduate School)

Leemis, Lawrence M. (1984), Stochastic Lifetimes: A General Model. (College of William & Mary)

Song, Whey-Ming (1988), Estimators of the Variance of the Sample Mean: Quadratic Forms, Optimal Batch Sizes, and Linear Combinations.

Avramidis, Athanassios (1993), Variance Reduction Techniques for Simulation with Applications to Stochastic Networks.

Hashem, Sherif (1993), Optimal Linear Combinations of Neural Networks. (University of Cairo)

Chen, Huifen (1994), Stochastic Root Finding in System Design. (Chung Yuan Christian University)

Pedrosa, Antonio (1994), Automatic Batching in Simulation Output Analysis. (Instituto Superior Técnic, Portugal)

Wang, Jin (1994), Contributions to Monte Carlo Analysis: Variance Reduction, Random Search, and Bayesian Robustness. (Valdosta State University)

Ceylan-Wood, Demet (1995), Variances and Quintiles in Dynamic-System Performance: Point Estimation and Standard Errors.

Jin, Jihong (1998), Simulation-Based Retrospective Optimization of Stochastic Systems.

Giddings, Angela (2002), A Unified Approach to Statistical Assessment of Heuristic Quality in Combinatorial Optimization.

Yeh, Yingchieh (2002), Steady-State Simulation Output Analysis: MSE-Optimal Dynamic Batch Means with Parsimonious Storage.

Pasupathy, Raghubhushan K. (2005), Retrospective-Approximation Algorithms for the Multidimensional Stochastic Root-Finding Problem. (Virginia Tech)

Supervised by Prof. Kirkpatrick (Case Institute):

Hwang, Sheue-Ling (1984), Human Supervisory Performance in Flexible Manufacturing.

Supervised by Prof. Taaffe (Minnesota):

Ong, Kim L. K. (1985), Approximating Nonstationary Multivariate Queueing Models.

Johnson, Mary (1988), Phase Distributions: Selecting Parameters to Match Moments.

Supervised by Prof. Wilson (Texas, NC State):

Tew, Jeff (1986), Metamodel Estimation Under Correlation Methods for Simulation Experiments. (Penn State)

Bauer, Kenneth (1987), Control Variate Selection for Multiresponse Simulation.

Venkatraman, Sekhar (1988), Modeling Multivariate Populations with Translation Systems.

Avramidis, Athanassios (1993), Variance Reduction Techniques for Simulation with Applications to Stochastic Networks.

Flanigan, Mary A. (1993), A Flexible, Interactive, Graphical Approach to Modeling Stochastic Input Processes.

Supervised by Prof. Rardin (Georgia Tech, Arkansas):

Campbell, Brian (1987), Steiner Tree Problems on Special Planar Graphs.

Sudit, Moises (1988), Paroids: A Generic Environment for Local Search. (SUNY Buffalo)

Mooney, Ed (1991), Tabu Search Heuristics for Resource Scheduling with Course Scheduling Applications. (Montana State University)

Ng, Peh (1991), Leontief Flow Problems: Integrality Properties and Strong Extended Formulations. (University of Minnesota)

Rais, Abdur (1992), The 2-Connected Steiner Subgraph Problem.

Morris, Sarah (1995), Simultaneous Wide-Area and Local-Access Network Design.

Venkatadri, Uday (1995), Fractal Layout for Job Shops.

Ghashghai, Elham (1997), Hybrid of Exact and Genetic Algorithms for Graph Optimal Problems with Application to Survivable Network.

Walter, Joseph (2001), Modeling of Free Flight.

Asmundsson, Jakob (2003), Tractable Nonlinear Capacity Models for Aggregate Production Planning.

Preciado-Walters, Felisa (2003), Optimal External Radiation Therapy Planning for Cancer: A Mixed Integer Approach.

Naidoo, Ramu (2004), Planning Models for Electric Transmission Network Expansion.

Supervised by Prof. Wagner (Northwestern):

DelGreco, John (1988), Representations of Bicircular Matroids and the Complexity of Recognizing a Class of Generalized Network Flow Matrices.

Gardner, Leslie (1989), A Decomposition of 3-Connected Graphs with Decomposition-Based Optimization Algorithms. (University of Indianapolis)

Rais, Abdur (1992), The 2-Connected Steiner Subgraph Problem.

Supervised by Prof. Chandru (MIT, Banagalore, India).

Dutta, Debasish (1989), Variable Radius Blends and Dupin Cyclides. (University of Michigan)

Aman, Amril (1992), On-Line Scheduling and Dynamic Task Assignment. (Jurusan Matematika Universitas, Indonesia)

Supervised by Prof. Coullard (Northwestern):

Gardner, Leslie (1989), A Decomposition of 3-Connected Graphs with Decomposition-Based Optimization Algorithms. (University of Indianapolis)

Ng, Peh (1991), Leontief Flow Problems: Integrality Properties and Strong Extended Formulations. (University of Minnesota)

Supervised by Prof. Shaw (Columbia):

Cho, Geon (1994), Limited Column Generation and Related Methods for Local Access Telecommunication Network Design and Expansion-Formulation, Algorithm, and Implementation.

Chang, Hsuliang (1998), Column Generation Approach for the Traveling Salesman Problem.

Supervised by Prof. Prabhu (NYU Courant Inst, MIT):

Zhang, Zaili (1995), Topics in Linear, Dynamic and Multi-Objective Optimization.

DeGuzman, Maria (1998), Experiments with Nonlinear Discriminants in the Analysis of Fine Needle Aspirates.

He, Wei (2002), New Methods for Computation of Zeros of Smooth Functions and Eigen-Decompositon of Symmetric Matrices.

Supervised by Prof. Papastavrou (MIT):

Kleywegt, Anton (1996), Dynamic and Stochastic Models with Freight Distribution Applications. (Georgia Tech)

Chang, H. C. (2000), Optimization on Nonlinear Surfaces.

Supervised by Prof. O'Cinneide (Cork, Ireland, Kentucky):

Warren, Graeme (1997), Analysis of Some Fluid Models and a Queueing Network Analyzer for Polling Systems.

Chen, Wu-Lin (1999), Transform Inversion and Its Application to Stochastic Models.

Kim, Sunkyo (2000), Refined Parametric Decomposition Approximation of Queueing Networks.

Appendix 10B | Ph.D.s in Manufacturing

Supervised by Prof. Lascoe (Western Kentucky):

Lin, Ian B. (1961), A Test Method for the Measurement and Evaluation of Shrinkage Characteristics of Epoxy Formulations During the Curing Cycle.

Packer, Kenneth F. (1962), A Study of Backward Extrusion of Steels at Slightly Elevated Temperatures.

ElGomayel, Joseph I. (1963), Tool Wear Relationships for Ceramic Cutting Tool Materials. (Purdue University)

Thomas, Stanislaus S. (1967), Certain Factors of Optimal Influence in Electric Discharge Machining. (University of Newark, New Jersey)

Dizer, John (1969), Tool Design Variables: Their Influence on the ECM Process.

Shalaby, El-Said E. (1970), Evaluation and Analysis of Cast High Speed Steel Cutting Tools. (University of Cairo)

Taroepratjeka, Harsono (1970), Investigation of the Pressure Behavior and Reflector Effects in an Electrohydraulic Process. (University of Bandung, Indonesia)

Fadhli, Atir (1972), The Potential of Ultrasonics in Shearing Micro-Size Sheet Metal.

Wu, Jack (1973), Analytical Model of Curriculum Development & Evaluation for Mfg. Engr. Education Programs.

Tseng, Mitchell M. (1975), A Systematic Approach to the Adaptive Control of the Electro-discharge Machining Process. (Hong Kong University of Science and Technology)

Reda, Mostafa (1976), Electro-Discharge Machining: The Influence of the Recast Layer and the Heat Affected Zone on Fatigue Strength.

Chui, Kwok-Sang (1978), Investigation on the Effect the Strain Ration has on Sheet Metal Formability.

Supervised by Prof. Barash (Technion, Israel, Manchester):

Wager, John G. (1967), The Nature and Significance of the Distribution of High-Speed-Steel Tool Life. (University of South Wales, Australia)

Berra, Peter B. (1968), Investigation of Automated Planning and Optimization of Metal Working Processes.

David, Larry G. (1968), A Study of Residual Stresses by High Energy Electric Discharge Machining.

Watanapongse, Dhani (1969), Investigation of the Shape of Sheets in Cold Rolling and a Method of Control by Computer.

Schoech, William J. (1971), A Study of the Residual Stress Distribution in Machining and a Semi-Analytical Model of the Plastically Deformed Zone. (Valparaiso University, Indiana)

Tomko, George M., Jr. (1971), The Effect of Residual Stresses Confined in a Thin Layer of Metal.

Weinstein, Jeremy S. (1971), Computer Control of Steel Tubemaking.

Batra, Jawahar L. (1972), Computer-Aided Planning of Optimal Machining Operations for Multiple-Tool Setups with Probabilistic Tool Life.

Raley, Frank (1972), The Effect of Residual Stress on the Impact Strength of Steel. (University of North Dakota)

Liu, Chunghorng (1973), An Investigation of the Surface Condition in a Steel Component Machined by Chip Removal Process. (Purdue University)

Hsu, John-Ping (1974), A System Modeling, Simulation and Control Study of the Steel Strip Cooling Process on the Runout Table of a Hot Rolling Mill.

Kotval, Cyrus J. (1975), Investigation of the Effect of Changing the Thermal Conductivity of Metal Cutting Tools on Their Wear.

Gupta, Surendra M. (1977), Computer-Aided Selection of Machining Cycles and Cutting Conditions on Multi-Stations Synchronous Machines. (Northeastern University)

Lewis, William C., Jr. (1981), Data Flow Architectures for Distributed Control of Computer Operated Manufacturing Systems: Structure and Simulated Applications.

Choi, Byoung K. (1982), CAD/CAM Compatible Tool-Oriented Process Planning for Machining Centers. (Korean Adv. Inst. of Technology)

Matsumoto, Yoichi (1983), Mechanics of Chip Formation and Its Effect on the Surface Integrity of Hardened Steels.

Donmez, M. Alkan (1985), A General Methodology for Machine Tool Accuracy Enhancement: Theory, Application and Implementation.

Riesser, William F. (1985), Electrotribology in Metal Cutting.

Venugopal, Raghunath (1985), Thermal Effects on the Accuracy of Numerically Controlled Machine Tools.

Wu, Herng-Liang (1985), Computer-Aided Configuration of Flexible Manufacturing Systems.

Eshel, Gad (1986), Automatic Generation of Process Outlines of Forming and Machining Processes.

Lee, Suk-Ho (1986), Accuracy Improvement of a CNC Machining Center by using a Touch Probe and a Metrology Pallet.

Hsu, Danny (1987), A New Approach to Select Strategy for Developing Countries to Build a Machine Tool Industry.

Chou, Yon-Chun (1988), Automatic Design of Fixtures for Machining Processes. (University of Massachusetts)

Lang, George (1988), Aid for the User of a CAD System: CAD-COACH–The Expert Tutor.

Upton, David M. (1988), The Operation of Large Computer-Controlled Manufacturing Systems. (Harvard University)

Ouyang, Yew-Shing (1989), Bimetal Forming Mechanics with Special Reference to Indentation, Extrusion and Upsetting.

Eneyo, Emmanual (1991), An Integrated Knowledge-Based Approach to Maintenance Control Systems for Automated Manufacturing.

Jahn, Chungen (1991), Concurrent Design for Constraint-Oriented Permanent Assembly.

Veeramani, Raj (1991), Physical Resource Management in Large Computer- Controlled Manufacturing Systems.

Kim, Kang (1992), Cylindricity Control in Precision Centerless Grinding.

Deshmukh, Abhijit (1993), Complexity and Chaos in Manufacturing Systems. (Texas A&M)

Chou, Yuag-Shan (1994), Wear Mechanisms of Cubic Boron Nitride Tools in Precision Turning of Hardened Steels.

Yang, Erh-Huan (1994), Operation Strategies for Large Computerized Manufacturing Systems.

Ahmed, Shahid (1997), A Concurrent Fixture Design and Assembly Methodology for Computer Integrated Manufacturing Systems.

Supervised by Prof. ElGomayel (Purdue):

Abou-Zeid, Mohammad R. A. (1973), SAGT–A New Coding System and the Systematic Analysis of Metal Cutting Operations. (University of Georgia)

Zakaria, Adel (1973), On-Line Tool Wear Sensing for Adaptive Control of Metal Cutting Processes.

Phillips, Rohan (1978), A Computerized Process Planning System Based on Component Classification and Coding.

Bregger, Klaus D. (1979), On-Line Tool Wear Sensing for Turning Operations.

Leep, Herman R. (1979), Evaluation of Synthetic Cutting Fluids in Metal Cutting Operations. (University of Louisville)

Philipson, Roland H. (1979), An Algorithm for the Optimization of a Multiple-Spindle Automatic Bar Machine.

Nader-Esfahani, Vahid (1980), Optimization of Machine Setup and Tooling Using Principles of Group Technology.

Roushdy, Essam H. (1980), CERMET Cutting Tools, Properties and Prediction of Performance.

Pugh, Gardner A. (1982), Inspector Allocation in a Production Environment. (Indiana University-Purdue University at Indianapolis)

Sanii, Ezatollah (1982), A Computerized Process Planning System Using Tool Classification and Coding. (North Carolina State University)

Supervised by Prof. Liu (Purdue):

Wu, Der-Jiunn (1982), A Mathematical Model of Machining Chatter.

Matsumoto, Yoichi (1983), Mechanics of Chip Formation and Its Effect on the Surface Integrity of Hardened Steels.

Lin, Zone-Ching (1984), A Quasi-Steady State Thermo-Elasto-Plastic Analysis of Stress Distribution in the Workpiece in Machining. (National Inst. of Science and Technology, Taiwan)

Donmez, M. Alkan (1985), A General Methodology for Machine Tool Accuracy Enhancement: Theory, Application and Implementation.

Lin, Yhu-Tin (1985), Geometric Adaptive Control for Accuracy and Stability in Machining Cylindrical Workpiece.

Srinivasan, Ramesh (1986), A Generalized Analytical Methodology for Generative Process Planning.

Ferreira, Placid (1987), Adoptive Accuracy Improvement of Machine Tools.

Liang, Gau R. (1987), Logic Approach to Surface-Generating Problem.

Wu, Muh-Cherng (1988), A New Methodology for Automatic Process Planning and Execution Based on Adaptive Information Modeling. (National Chiao Tung University, Taiwan)

Chao, Ping-Yi (1989), Determination of Manufacturability and Machining Parameters Based on a New Classification Method for Machinability. (Sun Yat-Sen University, Taiwan)

Roy, Utpal (1989), Computer Aided Representation and Analysis of Geometric Tolerances. (Syracuse University)

Trappey, Jui-Fen (1989), Methodology for Automatic Fixture Design in Computer Integrated Environment. (Tsinghua University, Taiwan)

Jan, Hung-Kang (1992), Dynamic Modeling of Manufacturing Process Error Patterns Using Distributed Adaptive Systems. (Yunlin University, Taiwan)

Kreng, Bor-Wen (1992), Intelligent Knowledge Management Environment for Design and Manufacturing. (Chung Hua University, Taiwan)

Mou, Jong-I (1992), An Adaptive Methodology for Monitoring and Controlling of Precision Machining and On-Machine Inspection. (Arizona State University)

Jang, Jaejin (1993), The Operation and Evaluation of Flexible Manufacturing Systems.

Mukherjee, Amit (1995), A Methodology for Representation and Reasoning in Conceptual Design: Application in Creating an Integrated Design Environment for Stamped Metal Parts

Mittal, Shridhar (1996), Feasibility of Single Step Superfinish Hard Machining and Its Effect on Surface Integrity.

Ahmed, Shahid (1997), A Concurrent Fixture Design and Assembly Methodology for Computer Integrated Manufacturing Systems.

Wang, Jia-Yeh (1998), A New Methodology for Analyzing The Heat Transfer and Thermal Damage Considering Tool Flank Wear in Finish Hard Machining.

Agha, Salah R. (2000), Fatigue Performance of Superfinish Hard Turned Surfaces in Rolling Contact.

Guo, Yuebin (2000), Finite Element Analysis of Superfinished Hard Turning.

Yang, Xiaoping (2001), A Methodology for Predicting the Variance of Fatigue Life Incorporating the Effects of Manufacturing Processes.

Byun, Jeongmin (2003), Methods for Improving Chucking Accuracy of a Cylindrical Workpiece in Finish Hard Turning.

Shi, Jing (2004), Prediction of Thermal Damage in Superfinish Hard Machined Surfaces.

Supervised by Prof. Chang (Virginia Tech):

Joshi, Sanjay (1987), CAD Interface for Automated Process Planning. (Penn State)

Irizarry-Lopez, Vilma (1989), A Methodology for the Automatic Generation of Process Plans in an Electronic Assembly Environment.

Lin, Alan (1990), Automated Assembly Planning for Three-Dimensional Mechanical Products. (Taiwan National University)

Wang, Ming T. (1990), A Geometric Reasoning Methodology for Manufacturing Feature Extraction from a 3-D CAD Model. (Taiwan National University)

Joneja, Ajay (1993), Automatic Design of Fixture Configurations: Representation and Planning. (Hong Kong University of Science and Technology)

Lee, Yuan-Shin (1993), Automatic Planning and Programming for Five-Axis Sculptured Surface Machining.

Yut, Greg (1996), Process Plan Feasibility: Accounting for Interactions Between Planning Decisions.

Hwang, Ji S. (1997), Five-Axis NC Machining of Compound Sculptured Surfaces.

Suh, Byung-Gyo (2000), Disassembly Planning with Consideration of Rotational Motions.

Hong, Yoo-Suk (2002), Tolerancing Algebra for Design and Manufacturing. (University of Toledo)

Supervised by Prof. Chandrasekar (Arizona):

Ahn, Yoomin (1992), Deformation about Sliding Indentation in Ceramics and Its Application to Fine Finishing.

Hebbar, Rajadasa (1992), Micro-Hole Drilling by Electrical Discharge Machining.

Chen, Yen-Meng (1994), The Cutting of Brittle Materials Using the Precision Crack-Off Process.

Xu, Yanwu (1994), Numerical Simulation and Theoretical Analysis of Plane-Strain Compression between Perfectly Rough Dies.

Lu, Ling (1995), Formation of Wear Particles in Polishing of Brittle Solids and Grinding of Metals.

Taylor, James B. (1995), A Precision Micro-Positioner for Manufacturing and Metrology. (Rochester Inst. of Technology)

Madhavan, Viswanathan (1996), Investigations into the Mechanics of Metal Cutting. (Wichita State University)

Swain, Selden (1996), Finish Machining of Hardened Steels Using CBN Cutting Tools.

Bulsara, Vispi H. (1997), Scratch Formation in Brittle Solids and Its Application to Polishing.

Ackroyd, Ben (2000), A Study of the Chip-Tool Contact Conditions in Machining.

Chhabra, Paul (2000), A Study of Adhesion and Friction in Intimate Metal-Metal Contacts.

Kompella, Sridhar (2002), Thermal Effects in Grinding.

Hwang, Jihong (2005), Contact Conditions at the Chip-Tool Interface In Machining.

Moylan, Shawn (2006), High-Speed Micro-Electro-Discharge Machining.

Shankar, M. Ravi (2006), High-Strength, Thermally-Stable Nano-structured Materials. (University of Pittsburgh)

Supervised by Prof. Chu (MIT, Korean Inst. of Tech.):

Kim, Kang (1992), Cylindricity Control in Precision Centerless Grinding.

Supervised by Prof. Tu (Michigan, NC State):

Lankalapalli, Kishore (1996), Model-Based Penetration Depth Estimation of Laser Welding Processes.

Bossmanns, Bernd (1997), Thermo-Mechanical Modeling of Motorized Spindle Systems for High Speed Milling.

Katter, James (1997), Condition Monitoring of High Power CO_2 Laser Welder Systems for Preventative Maintenance.

Cho, Nam (2001), Circularity Tolerance Modeling, Analysis, and Design for High Precision Assemblies.

Lin, Chi-Wei (2001), High Speed Effects and Dynamic Analysis of Motorized Spindles for High Speed End Milling.

Supervised by Prof. Compton (Illinois):

McComb, Sara A. (1998), The Effects of Dynamic Team Behavior and Task Complexity on Team Performance. (Texas A&M)

Ackroyd, Ben (2000), A Study of the Chip-Tool Contact Conditions in Machining.

Chhabra, Paul (2000), A Study of Adhesion and Friction in Intimate Metal-Metal Contacts.

Monahan, David (2003), Analysis and Quantification of Technological Cycles in High Technology Industries.

Swamingman, S (2006), Nanoscale Microstructures in Substitutional Solid Solutions by Large Strain Machining.

Shankar, M. Ravi (2006), High-Strength, Thermally-Stable Nanostructured Material. (University of Pittsburgh)

Moscuso, W. (2008), Severe Plastic Deformation and Nanostructured Materials by Large Strain Extrusion Machining.

APPENDIX 10C | Ph.D.s in Human Factors

Supervised by Prof. Mundel (NYU, Iowa, Bradley, Maryland):

Thomas, Orpha M. H. (1946), A Scientific Basis for the Design of Institution Kitchens.

Lehrer, Robert N. (1949), Development and Evaluation of a Pace Scale for Time Study Rating. (Northwestern University)

Nadler, Gerald (1949), A Priority Procedure for Biomechanical Determination of Optimum Motion Pattern. (University of Southern California)

Supervised by Prof. Richardson (Lehigh):

Lazarus, Irwin P. (1952), A System of Predetermined Human Work Times.

Wimmert, Robert J. (1957), A Quantitative Approach to Equipment Location in Intermittent Manufacturing. (Lehigh University)

Supervised by Prof. Amrine (Ohio State):

Gambrell, Carroll Blake Jr. (1958), The Relationship Between Time Study Pace Rating and Job Difficulty. (Mercer University)

Nichols, D. Edward (1958), Physiological Evaluation of Selected Principles of Motion Economy. (University of Rhode Island)

Barany, James W. (1961), The Nature of Individual Differences in Bodily Force Exerted During a Motor Task as Measured by a Force-Platform.

Hoyt, Charles D. (1962), Labor Productivity as a Basis for Variable Budget Control of Manufacturing Operations. (Auburn University)

Supervised by Prof. Greene (Iowa):

Glickstein, Aaron (1960), The Development of an Integrated Production Control System Through Simulation Procedures.

Barany, James W. (1961), The Nature of Individual Differences in Bodily Force Exerted During a Motor Task as Measured by a Force-Platform. (Purdue University)

James, Charles F., Jr. (1963), The Development of an Analytical and Predictive Technique for Industrial Wage Structure. (Univ. of Milwaukee)

Mann, Lawrence, Jr. (1965), Development of a Procedure for Predicting Roadway Maintenance Costs. (Louisiana State University)

Scheck, Donald E. (1966), Feasibility of Automated Process Planning.

Sharp, James F. (1966), Nonlinear, Multifacility Production-Transportation-Marketing Models: Analysis and Algorithms.

Bakr, Mamdouh M. (1968), Operating Control in Continued Production Systems.

Puscheck, Herbert C. (1969), The Development and Application of a Simple Wargame to the Study of Sequential Decision Making in a Conflict Environment.

Supervised by Prof. Ritchey (Univ. of Chicago):

Huber, George P. (1966), A Systematic Approach to the Industrial Staffing of Professionals. (University of Texas)

Supervised by Prof. Barany (Purdue):

Nelson, J. Byron (1969), A Dynamic Visual-Recognition Test for Paced Inspection Tasks.

Smith, Leo A. (1969), Analysis of Errors on Paced Visual Inspection Tasks.

Bohlen, George (1973), A Learning Curve Prediction Model for Operators Performing Industrial Bench Assembly Operations. (University of Dayton)

Hosein, Robert W. (1973), The Effect of Operator Performance Variability on Assembly Line Balancing.

Benysh, Susan (1998), The Effects of Knowledge Type and Warning Format on Attention to Elements within Instructional Material and Warning Compliance.

Supervised by Prof. Buck (Michigan, Iowa):

Rizzi, Anthony M. (1971), A Performance Model of 'The Paced Visual Inspection Task' With Analysis of Four Paced Task Combinations and Their Associated Eye-Motion Patterns.

Lee, David R. (1972), An Evaluation of Selected Parameters at the Microform Reading Man-Machine Interface. (University of Dayton)

Tanchoco, Jose M. (1975), Transform Methods for an Economic Analysis of Probalistic Cash Flows. (Purdue University)

Wentworth, Ronald N. (1981), Eye Movement Behavior as a Predictor of Performance on Paced Visual Inspection Tasks.

Supervised by Prof. Salvendy (Birmingham, Beijing):

Joost, Michael G. (1976), Quantification of Hyperactivity in Children.

Boydstun, Louis E. (1977), Hierarchical Acquisition of Psychomotor Skills.

Suominen, Satu M. (1982), Impact of Changes in Physical Fitness on the Effectiveness of Decision Making.

Sharit, Joseph (1984), Human Supervisory Control of a Flexible Manufacturing System: An Exploratory Investigation. (University of Buffalo)

Barfield, Woodrow (1986), Cognitive and Perceptual Aspects of Three Dimensional Figure Rotations for Computer-Aided Design (CAD) Systems.

Garg, Chaya (1987), Development of a Methodology for Knowledge Elicitation for Building Expert Systems.

Koubek, Richard (1987), Toward an Understanding of Super-Expert Cognitive Performance: Implications for Expert Systems and Software Engineering. (Penn State)

Gibson, David (1989), Knowledge Structures in Human Problem Solving: Implications for Human-Interactive Tasks.

Papantonopoulos, S.A. (1990), A Decision Model for Cognitive Task Allocation.

Cook, John R. (1991), Cognitive and Social Factors in the Design of Computerized Jobs.

Ye, Nong (1991), Development and Validation of a Cognitive Model of Human Knowledge Systems: Toward an Effective Adaptation to Differences in Cognitive Skills. (University of Central Florida)

Chao, Chin-Jung (1992), Development of a Methodology for Optimizing the Elicited Knowledge.

Stanney, Kay (1992), Effects of Diversity in Field-Articulation on Human-Computer Performance.

Zhao, Baijun (1992), A Structured Analysis and Quantitative Measurement of Task Complexity in Human-Computer Interaction.

Bi, Shuxin (1993), Workload Prediction in the Design of Dynamic Control Systems: Application to Manufacturing Systems.

Jacko, Julie A. (1993), Models for Hierarchical Menu Design: Incorporating Breadth and Depth, Task Complexity, and Knowledge Structure of the User. (Georgia Tech)

Gong, Qing (1994), Development and Validation of a Conceptual Model for a Skill-Based Adaptive Human-Computer Interface.

Patel, Umesh (1995), An Analog of Ohm's Law for Modeling Mental Workload.

Choong, Yee-Yin (1996), Design of Computer Interfaces for the Chinese Population.

Duffy, Vincent (1996), Development and Validation of a Model for Successful Integration of People, Organization and Technology in Concurrent Engineering: A Study of 103 Electronics Manufacturers. (Purdue University)

Matias, Aura (1996), Predictive Models of Carpal Tunnel Syndrome (CTS) Causation among VDT Operators. (University of the Philippines)

Dong, Jianming (1997), Human-Computer Interface Design for the Chinese Population.

Lin, Xianzhan (1997), Development and Validation of Cognitive Models for Human Error Reduction.

Xie, Bin (1997), A Methodology for Modeling and Predicting Mental Workload in Single and Multi-Task Environments.

Chen, Bo (1998), Design and Validation of a User-Centered Web Browser.

Rau, Pei-Luen (1998), Design and Evaluation of a Human-Centered Electronic Mail Address System.

Fang, Xiaowen (1999), Development and Validation of a User-Centered Search Tool for the Web.

Fu, Limin (1999), Usability Evaluation of Web Page Design.

McKinnis, David R. (1999), The Role of Faculty in University-Based Industrial Outreach. (Purdue University)

Liu, Baili (2000), Modeling the Design of Visual Search Tasks in Human-Computer Interaction.

Wei, June (2000), The Cognitive Task Analysis System.

Ji, Yong G. (2001), Intranet Portal Organizational Memory Information Systems.

Ozok, Ant (2001), Guidelines for the Design and Measurement of Interface Consistency in the World Wide Web.

Ma, Lingfeng (2002), Search Tools for Internet2.

Xie, Yunlong (2002), Situation Awareness Support for Asynchronous Engineering Collaboration.

Song, Guangfeng (2003), A Method for Reusing Web Browsing Experience to Enhance Web Information Retrieval. (Penn State)

Horn, Diana (2005), Modeling and Quantifying Consumer Perception of Product Creativity.

Liu, Yan (2006), Interactive Visual Data Mining Modeling to Enhance Understanding and Effectiveness of the Process.

Supervised by Prof. Lehto (Michigan):

Foley, James P. (1990), The Effect of Law and Training on All-Terrain Vehicle Riders' Safety-Related Behaviors.

Calisir, Fethi (1996), Human Decision Making and Seat Belt Use.

Chatterjee, Subrata (1998), A Connectionist Approach for Classifying Accident Narratives.

Liang, Sheau-Farn (1999), Information Representation and Decision Process: Effects of Measurement Scale and Shape of Decision Matrix on Preferential Choice.

Yang, Yanxia (2001), Graphical Information for Blind Users of the Web.

Supervised by Prof. Eberts (Illinois):

Bringelson, Liwana (1991), Group Mental Model Transfer.

Posey, Jack (1992), Pictorial and Text Editors for Expert System Rules.

Villegas, Leticia (1993), User Modeling in Human-Computer Interaction Tasks.

Phillips, Colleen L. (1998), Intelligent Support for Engineering Collaboration.

Supervised by Prof. Koubek (Penn State):

Tang, Kuo-Hao (1995), Development and Validation of a Theoretical Model for Cognitive Skills Acquisition.

Harvey, Craig (1997), Toward a Model of Distributed Engineering Collaboration. (University of Oklahoma)

Benysh, Darel (1998), Development of a Theoretical Framework and Design Tool for Process Usability Assessment.

Appendix 10D | Ph.D.s in Systems Engineering

Supervised by Prof. Owen:

Ayasun, Nurettin (1944), Design, Location, Organization, and Industrial Management of a Munitions Plant.

Supervised by Prof. Young (Arizona State):

Bedworth, David D. (1961), Design of An Electro-Mechanical Device for Simulating Product Flow Characteristics in a Manufacturing Line. (Arizona State University)

Mize, Joe H. (1964), A Heuristic Scheduling Model for Multi-Project Organizations. (Oklahoma State University)

Moodie, Colin L. (1964), A Heuristic Method of Assembly Line Balancing for Assumptions of Constant or Variable Work-Element Times. (Purdue University)

Moreno, Carlos W. (1966), A Technique for Simulating Transient Sequential Queues in Production Lines.

Supervised by Prof. Brooks (Georgia Tech):

Beenhakker, Henri L. (1963), The Development of Alternate Criteria for Optimality in the Machine Sequencing Problem.

Thomas, Warren H. (1964), Model for Short Term Prediction of Demand for Nursing Resources. (SUNY Buffalo)

Supervised by Professor R. Reed (Georgia Tech):

Cusack, George M. (1968), An Investigation of the Job Shop Loading Problem.

Roberts, Stephen D. (1968), Warehouse Size and Design. (North Carolina State)

Gibson, David F. (1969), Objective Assignment of Organizational Resources to Projects.

Urban, William W. (1969), Project Evaluation of Functionally Organized Industrial Engineering/Development.

Webster, Dennis B. (1969), Determination of a Materials Handling Systems Decision Model. (Auburn University)

Emerson, C. Robert (1970), A Parking Analysis and Planning Technique. (SUNY Binghamton)

Sapp, Richard S. (1971), The Procurement Process and Program Cost Outcomes: A Systems Approach.

Bafna, Kailash M. (1972), An Analytical Approach to Design High-Rise Stacker Crane Warehouse Systems. (Western Michigan University)

Somers, James L. (1972), PACE: A Methodology for Planning and Control of Job Shops.

Heisterberg, Rodney J. (1975), LOTSA: An Algorithm for Local Truckload Operations Control.

Shunk, Dan L. (1976), The Measurement of the Effects of Group Technology by Simulation. (Arizona State University)

Supervised by Prof. Moodie (Purdue):

Anderson, David R. (1968), Transient and Steady-State Minimum Cost In-Process Inventory Capacities for Production Lines. (University of Cincinnati)

Carlin, Jerry L. (1970), An Investigation of Some Public Security Allocation and Patrol Methods.

Mason, A. Thomas (1970), An Algorithm for a Costs Tradeoff Problem in Project Scheduling with a Single Resource.

Millen, Roger N. (1970), An Analysis of the Effects of Computer Process Control on Management Information and Control Systems.

Deane, Richard H. (1972), Scheduling Methodologies for Balancing Workload Assignments in the Job Shop Manufacturing Environment.

Jain, Amit K. (1972), A Study of Automating Managerial Decision Making and Its Effect on the Organization Structures.

Sadowski, Randall P. (1972), A Methodology to Aid in the Design of Unloading Facilities. (Purdue University)

Malstrom, Eric M. (1974), Adaptive Forecasting of Project Costs and Through-put Times in a Job Shop Production Facility.

Hebert, John E. (1975), Critical Path Analysis and a Simulation Program for Resource-Constrained Activity Scheduling in GERT Project Networks. (University of Arkansas)

Khator, Suresh (1975), A Loading Methodology for Job Shops having Conventional and NC Machine Tools.

Haider, Syed W. (1976), An Investigation of the Advantages of Using a Man-Computer Interactive Mode for Scheduling Job Shops.

Wysk, Richard A. (1977), An Automated Process Planning and Selection Program–APPAS. (Penn State)

Mackulak, Gerald T. (1979), A Production Control Strategy for Hierarchical Multi-Objective Scheduling with Specific Application to Steel Manufacture. (Arizona State)

Chen, Jhitang (1981), Integration of Process Planning with MRP and Capacity Planning for Better Shop Production Planning and Control.

Chu, Chi-Chung (1984), An Adaptive Decision Making Methodology for Material Handling Equipment in a Computer Integrated Manufacturing System.

Ben-Arieh, David (1985), Knowledge Based Control System for Automated Production and Assembly. (Kansas State University)

Lin, Chin-Wen (1985), An Energy-Effective Production Scheduling Strategy for Hierarchical Control of Steel Manufacture.

Kumara, Soundar (1985), Artificial Intelligence Techniques in Facilities Layout Planning Development of an Expert System.

Kacha, Pierre (1987), Development of a Frame-Based Model for Cut-Off Dimensioning.

Moon, Young Bai (1988), A Framework for Failure Recovery of Automated Manufacturing Cell Components.

Drolet, Jocelyn (1989), Scheduling Virtual Cellular Manufacturing Systems. (University of Quebec)

Banerjee, Pat (1990), A Manufacturing Layout Reasoning Architecture Based on an Automated Integration of Linear Objective Optimization and Non-Linear Qualitative Analysis. (University of Illinois)

Lee, Ching-En (1991), An Integrated Methodology for the Analysis and Design of Cellular Flexible Assembly Systems.

Webber, Detlef (1992), Information Management Architecture for an Advanced Integrated Manufacturing Enterprise.

Li, Hong (1994), A Formalization and Extension of the Purdue Enterprise Reference Architecture and the Purdue Methodology.

Ho, Ying-Chin (1995), Cell Formulation and Layout Design for Multiple-Cell Manufacturing with Flexible Processing and Routing.

Shewchuk, John P. (1995), A Systems Approach to Flexibility and Manufacturing Systems Analysis and Design. (Virginia Tech)

Warren, Graeme (1997), Analysis of Some Fluid Models and a Queuing Network Analyzer for Polling Systems. (University of Pretoria, South Africa)

Supervised by Prof. Pritsker (Ohio State, Arizona, Virginia Tech):

Hurst, Nicholas (1973), GASP IV: A Combined Continuous/Discrete FOR-TRAN Based Simulation Language.

Sigal, Charles E. (1977), The Stochastic Shortest Route Problem.

Standridge, Charles R. (1978), Incorporation of Database Systems Concepts into Simulation Modeling.

Wilson, James R. (1979), Variance Reduction Techniques for the Simulation of Queuing Networks. (North Carolina State University)

Bartkus, David E. (1981), A Dynamic Structural Equation Model of Physician Distribution.

Yancey, David P. (1981), A Framework for the Analysis of Transition Path Simulation Model. (University of North Carolina, Charlotte)

Bengston, Neal (1983), Development and Use of Operational Analysis Model Error Measures. (Barton College)

Supervised by Prof. Petersen (Arizona State):

Pegden, Claude D. (1975), An Implicit Enumerations Algorithm (GIPC2) for Solving Objective Functions. (Penn State)

Li, Ching-Ming (1983), An Integrated Production Planning and Control System for Steelmaking Facilities with an Energy Conservation Criterion".

Reilly, Charles (1983), PlP21: A Partial Enumeration Algorithm for Pure 0-1 Polynomial Integer Programming Problems.

Supervised by Prof. Talavage (Case Institute):

Davis, Wayne J. (1975), Three-Level Echelon Models for Organizational Coordination. (University of Illinois)

Fraley, David W. (1975), On the Identification and Control of Sequential Machines.

Musselman, Kenneth J. (1978), An Interactive, Tradeoff Cutting Plane Approach to Continuous and Discrete Multiple Objective Optimization.

Azidivar, Farhad (1979), Optimization of Stochastic Systems Through Simulation Using Stochastic Approximation Method. (University of Massachusetts, Dartmouth)

Fortuny-Amat, Jose (1979), Multi-Level Programming.

Chen, Peter H. C. (1986), A Model Formalism for the Design of Simulation Model Representative Languages.

Shodhan, Ronak H. (1989), COMAND: A Computer Consultant for the Design, Operations, and Control of Flexible Manufacturing Systems.

Chandra, Jayanta (1990), Optimization-Based Opportunistic Part Dispatching in Flexible Manufacturing Systems.

Benjaafar, Saifallah (1992), Modeling and Analysis of Flexibility in Manufacturing Systems.

Deshmukh, Abhijit (1993), Complexity and Chaos in Manufacturing Systems.

Tsai, Kune-Muh (1994), The Computation of the Variance of Throughput Rate of Production Systems from the Central Server Model.

Piplani, Rajesh (1995), Multi-Criteria Design and Control of Manufacturing Systems Using Simulation and Artificial Intelligence.

Supervised by Prof. Deisenroth (Georgia Tech, Virginia Tech):

Surh, Dae Suk (1979), Allocation of Files in a Hierarchical Information Control and Monitoring.

Gupta, Rajiv M. (1980), An Experimental Investigation of the Effects of Visual Aids in the Design of Industrial Facilities.

Supervised by Prof. Sadowski (Penn State):

Greene, Timothy J. (1980), Loading Concepts for the Cellular Manufacturing System. (Western Michigan University)

Sneider, Richard M. (1980), A Methodology for Optimal Assembly Line Balancing.

Medeiros, Deborah J. (1981), Scheduling Parallel Processors with Due Dates and Set Up. (Penn State)

John, Thuruthickara C. (1984), Studies in Multicriteria Scheduling Problems.

Harmonosky, Cathy (1987), An Approach to Generalized Analysis of Automated Manufacturing Systems through Classification. (Penn State)

Supervised by Prof. Roberts (Purdue, Florida, Indiana, NC State):

England, William L. (1982), Medical Diagnostic Test Sequencing and Optimal Protocol Design.

Flanigan, Mary A. (1993), A Flexible, Interactive, Graphical Approach to Modeling Stochastic Input Processes.

Supervised by Professor Nof (Michigan):

Fisher, Edward L. (1984), Knowledge-Based Facilities Design.

Maimon, Oded A. (1984), Activity Controller for a Multiple Robot Assembly Cell.

Rajan, Venkat (1993), Cooperation Requirement Planning for Multi-Robot Assembly Cells.

Witzerman, Jim (1995), A Facility Description Language for Distributed Design Integration.

Kim, Chang-ouk (1996), DAF-Net and Multi-Agent Based Integration Approach for Heterogeneous CIM Information Systems.

Williams, NaRaye (1998), The Effectiveness of Protocol Adaptability in TestLAN-Based Production Environments.

Ceroni, Jose (1999), Models for Integration with Parallelism of Distributed Organizations.

Huang, Chin-Yin (1999), Autonomy and Viability in Agent-Based Manufacturing Systems.

Lara Gracia, Marco (1999), Conflict Resolution in Collaborative Facility Design.

Chen, Jianhao (2002), Modeling and Analysis of Coordination for Multienterprise Networks.

Anussornnitisarn, Pornthep (2003), Design of Active Middleware Protocols for Coordination of Distributed Resources.

Supervised by Prof. Tanchoco (Purdue):

Occena, Luis (1987), A Pattern Directed Inference Approach to Hardwood Log Breakdown Decision Automation.

Taghaboni, Fataneh (1989), Scheduling and Control of Manufacturing Systems with Critical Material Handling.

Kim, Chang (1991), The Operation of an Automated Guided Vehicle System in a Manufacturing Job Shop.

Noble, Jim (1991), A Framework for the Design Justification of Material Handling Systems.

Rembold, Bernhard (1992), An Integrated Framework for the Design of Material Flow Systems.

Sinriech, David (1993), A Design Framework Based on the Segmented Flow Topology (SFT) for Discrete Material Flow Systems.

Koo, Pyung-Hoi (1996), Flow Planning and Control of Single-Stage Multimachine Systems.

Lee, Andrew C. H. (1999), Distribution Network Logistics for After Sale Service.

Ting, Juu-Hwa (1999), Orthogonal Spatial Material Flow Structures with Application to 300mm Semiconductor FabAutomation.

Yang, Su-Hsia (1999), Decision Support System for Machine Re-Layout Planning.

Supervised by Prof. Yih (Wisconsin):

Hashem, Sherif (1993), Optimal Linear Combinations of Neural Networks.

Chiu, Chanshing (1994), A Learning-Based Methodology for Dynamic Scheduling in Distributed Manufacturing Systems.

Ge, Yang (1995), Crane Scheduling with Time Windows in Flowshop Environments.

Chang, Te-Min (1996), A Fuzzy Rule-Based Methodology for Dynamic Kanban Control in a Generic Kanban System.

Chen, Chun-Chung (1997), A Learning-Based Methodology for Scheduling Problems in Semiconductor Fabrication Plants.

Sun, Yu-Liang (1997), An Intelligent Controller for Manufacturing Cells.

Lu, Ta-Ping (2000), An Agent-Based Production Control Framework for Collaboration in Supply Chain Management.

Min, Hyeung-Sik (2002), Development of a Real Time Multi-Objective Scheduler for Semiconductor Fabrication System.

Supervised by Prof. Uzsoy (Florida, NC State):

Ovacik, Irfan (1994), A Decomposition Methodology for Scheduling Complex Job Shops.

Mehta, Sanjay (1996), Predictable Shop Scheduling in the Presence of Machine Breakdowns.

Emirkol, Ebru (1999), Design and Development of Effective Decomposition Methods for Scheduling Complex Workshops.

Wang, Cheng-Shuo (2000), Decomposition Heuristics for Complex Job Shop Scheduling.

Akcali, Elif (2001), Tool and Setup-Constrained Capacity-Allocation Problem with Assignment Restrictions.

Geiger, Chris (2001), A Scheduling Rule Discovery and Parallel Learning System.

Giddings, Angela (2002), A Unified Approach to Statistical Assessment of Heuristic Quality in Combinatorial Optimization.

Asmundsson, Jakob (2003), Tractable Nonlinear Capacity Models for Aggregate Production Planning.

Supervised by Prof. Lawley (Illinois):

Gebraeel, Nagi (2003), Confidential Thesis.

Sulistyono, Widodo (2004), Deadlock Avoidance in Automated Manufacturing Systems With Unreliable Resource and Flexible Process Sequencing.

Notes

1. National Academy of Engineering, *The Engineer of 2020*, 57.
2. Salvendy, *Handbook of Industrial Engineering*, iv-vii.
3. Bush, *Science*, 34.
4. National Science Foundation, *Universities Report*, 1.
5. National Science Foundation Website. 2009. Directorate for Engineering, Civil, Mechanical and Manufacturing Innovation. http://www.nsf.gov/div/index.jsp?div=cmmi.
6. National Academy of Engineering, *Educating the Engineer of 2020*, 5.
7. Friedman, *The World is Flat; Hot, Flat, and Crowded*.
8. National Science Foundation, *Universities Report*, 25.
9. Leonhardt, "The Big Fix," 27.
10. National Academy of Engineering, *Rising Above the Gathering Storm*, 25.
11. Ibid., 30.

BIBLIOGRAPHY

Ackoff, Russell L. 1971. "Some Notes on Planning." Presentation at a meeting of American Association for the Advancement of Science, Philadelphia, PA, Dec. 30, 1971.

Ackoff, Russell L. 1974. *Redesigning the Future: A Systems Approach to Societal Problems*. New York: Wiley.

Ackoff, Russell L. 1978. *The Art of Problem Solving*. New York: Wiley.

Ackoff, Russell L. 1981. "The Art and Science of Mess Management." *Interfaces* 1: 11-16.

Ackoff, Russell L. 1999. *Re-Creating the Corporation: A Design of Organizations for the 21st Century*. Oxford: Oxford University Press.

Ackoff, Russell L. and H. L. Addison. 2007. *Management F-Laws: How Organizations Really Work*. Axminster, UK: Triarchy Press.

Ackoff, Russell L. and Daniel Greenberg. 2008. *Turning Learning Right Side Up: Putting Education Back on Track*. Boston: Pearson Books.

Adams, John. 2008. *Hallelujah Junction: Composing an American Life*. New York: Farrar, Straus, and Giroux.

Alford, L. P. 1935. *Henry Laurence Gantt, Leader in Industry*. New York: ASME.

Amrine, Harold T., John A. Ritchey, and Oliver S. Hulley. 1957. *Manufacturing Organization and Management*. Englewood Cliffs, NJ: Prentice Hall.

Amrine, Harold T., John A. Ritchey, and Colin L. Moodie. 1987. *Manufacturing Organization and Management*. Englewood Cliffs, NJ: Prentice Hall.

Amrine, Harold T. 1985. *Industrial Engineering at Purdue University; The Roots and First Thirty Years*. West Lafayette, IN: Purdue School of Industrial Engineering.

Arendt, Hanna. 1958. *The Human Condition*. Chicago: University of Chicago Press.

Arnold, Horace L. and Fay L. Faurote. 1915. *Ford Methods and the Ford Shops*. New York: The Engineering Magazine Company.

ASME. 1912. *The Present State of the Art of Industrial Management*. New York: American Society of Mechanical Engineers.

Babbage, Charles. 1846. *On The Economy of Machinery and Manufactures*. London: John Murray.

Babbage, Charles. 1889. *Babbage's Calculating Engines; Being a Collection of Papers Relating to Them, Their History and Construction.* London: E. and F. N. Spon.

Barash, Moshe M. 1982. *"Computerized Manufacturing Systems for Discrete Products." Handbook of Industrial Engineering, 7.9.1-7.9.7.*

Barfield, Woodrow and Tom Furness III. 1995. *Virtual Environments and Advanced Interface Design.* Oxford: Oxford University Press.

Barfield, Woodrow and Thomas Caudell. 2001. *Fundamentals of Wearable Computers and Augumented Reality.* Barfield, NY: Lawrence Erlbaum Associates.

Barnes, Ralph. 1980. *Motion and Time Study: Design and Measurement of Work.* New York: Wiley.

Bedworth, David D., Mark Henderson and Philip Wolfe. 1991. *Computer-Integrated Design and Manufacturing.* New York: McGraw-Hill.

Bedworth, David D., James Riggs and Sabah Randhawa. 1996. *Engineering Economics.* New York: McGraw-Hill.

Billington, David P. 1986. "In Defense of Engineers." *The Wilson Quarterly.* New Year's Issue (1986): 86-97.

Black, J. T. 1978, "IE's Have Roots, Too." *Industrial Engineering* (May) 5: 22-29.

Bowditch, John and Clement Ramsland. 1961. *Voices of the Industrial Revolution.* Ann Arbor: University of Michigan Press.

Brandeis, Louis D. "Scientific Management and Railroads." *Engineering Magazine.* 1911. Reprint. Easton, PA: Hive Publishing, 1981.

Bush, Vannevar. 1945. *Science, the Endless Frontier.* Washington, DC: National Science Foundation.

Chandrasekar, S., ed. 1988. *Intersociety Symposium on Machining of Advanced Ceramic Materials and Components.* New York: American Society of Mechanical Engineers.

Chandru, Vijay and John Hooker. 1999. *Optimization Methods for Logical Inference.* New York: Wiley.

Chang, Tien-Chien. 1983. *Advances in Computer-Aided Process Planning.* Washington, DC: National Bureau of Standards.

Chang, Tien-Chien, Richard A. Wysk, and Hsu-Pin Wang. 2006. *Computer-Aided Manufacturing.* Englewood Cliffs, NJ: Prentice Hall.

Chang, Tien-Chien. 1990. *Expert Process Planning for Manufacturing.* Boston: Addison-Wesley.

Chang, Tien-Chien and Richard A. Wysk. 1985. *An Introduction to Automated Process Planning Systems.* Englewood Cliffs, NJ: Prentice Hall.

Chang Tien-Chien, R. A. Wysk, and H. P. Wang. 1991. *Computer-Aided Manufacturing*. Englewood Cliffs, NJ: Prentice Hall.

Churchman, C. West, R. L. Ackoff, and E. L. Arnoff. 1957. *Introduction to Operations Research*. New York: John Wiley and Sons.

Churchman, C. West. 1968. *The Systems Approach*. New York: Delacorte.

Churchman, C. West. 1979. *The Systems Approach and Its Enemies*. New York: Basic Books.

Clark, Gregory. 2007. *A Farewell to Alms: A Brief Economic History of the World*. Princeton, NJ: Princeton University Press.

Compton W. Dale. 1997. *Engineering Management: Creating and Managing World Class Operations*. Englewood Cliffs, NJ: Prentice-Hall.

Compton, W. Dale and P. Reid Proctor. 2005. *Building a Better Delivery System: A New Engineering/Health Care Partnership*. Washington, DC: The National Academies Press.

Cooke, Morris L. and Philip Murray. 1940. *Organized Labor and Production*. New York: Harper.

Copley, Francis Barkley. 1923. *Frederick W. Taylor: Father of Scientific Management*. New York: Harper.

Coullard, Collette, Robert Fourer, and Jonathan H. Owen, eds. 2002. *Modeling Languages and Systems*. Boston: Kluwer.

Debris of 1899. 1899. Yearbook. West Lafayette, IN: Purdue University Class of 1899.

DeGarmo, E. Paul and B. M. Woods. 1942. *Introduction to Engineering Economy*. New York: Macmillan.

DeGarmo, E. Paul. 1969. "Industrial Engineering—Still Uniquely Defineable," *Industrial Engineering* (October): 70-71.

Deming, W. Edwards. 1986. *Out of the Crisis*. Cambridge: MIT Press.

Deming, W. Edwards. 2000. *The New Economics for Industry, Government, Education, 2nd Edition*. Cambridge: MIT Press.

Diemer, Hugo. 1925. *Factory Organization and Administration*. New York: McGraw-Hill.

Dillon, Mary, and Lillian Gilbreth. 1929. *Kitchen Practical: The Story of an Experiment*. Brooklyn, NY: Brooklyn Borough Gas Company.

Doherty, M. J. and F.T. Sparrow. 1983. *Mobility Enterprise One Year Later*. West Lafayette, IN: Purdue University Press.

Dorf, R. C. and S. Y. Nof, eds. 1990. *Concise International Encyclopedia of Robotics, Applications and Automation*. New York: John Wiley and Sons.

Drucker, Peter F. 1985. *Innovation and Entrepreneurship*. New York: Harper.

Drucker, Peter F. 1992. *Managing for the Future: The 1990s and Beyond*. New York: Dutton.

Drucker, Peter F. 1993. *Post-Capitalist Society*. New York: Harper.

Drucker, Peter F. 1995. *Managing in a Time of Great Change*. New York: Dutton.

Drucker, Peter F. 1999. *Management Challenge for the 21st Century*. New York: Harper.

Drucker, Peter F. 2002. *Managing in the Next Society*. New York: St. Martin's.

Drury, Horace B. 1915. *Scientific Management: a History and Criticism*. New York: Columbia Univesity Press.

Dubos, Rene. 1968. *So Human an Animal*. New York: Charles Scribner's Sons.

Duncan, Acheson J. 1974. *Quality Control and Industrial Statistics*. Homewood, IL: Irwin.

Eberts, Ray E. and Cindelyn Eberts, eds. 1985. *Trends in Ergonomics/ Human Factors II*. Amsterdam: North-Holland.

Eberts, Ray E. 1994. *User Interface Design*. Englewood Cliffs, NJ: Prentice Hall.

Eckles, Robert B. 1974. *The Dean: A Biography of A. A. Potter*. West Lafayette, IN: Purdue University Press.

Ellul, Jacques. 1948. *The Technological Society*. New York: Alfred A. Knopf.

Englund, Steven. 2008. "How Catholic is France." *Commonweal*, Nov. 7, p.12.

Emerson, Howard P. 1988. *Origins of Industrial Engineering*. Norcross, GA: Engineering and Management Press.

Evans, Gerald, Thomas Morin, and Herbert Moskowitz. 1980. *Uncertainty in Energy Generation Expansion*. West Lafayette, IN: Purdue University Press.

Feenberg, Andrew. 1996. *Heidegger, Habermas, and the Essence of Technology*. Kyoto: International Institute for Advanced Study.

Feenberg, Andrew. 1996. *Summary Remarks on My Approach to the Philosophic Study of Technology*. Palo Alto, CA: Xerox PARC.

Feenberg, Andrew. 2002. *Transforming Technology: A Critical Theory Revisited*. Oxford: Oxford University Press.

Florman, Samuel C. 1994. *The Existential Pleasures of Engineering*. New York: St. Martin's.

Ford, Wendy, Joanne Fox and Julie Sturgeon. 1999. *Regenstrief: Legacy of the Dishwasher King*. Indianapolis: Regenstrief Foundation.

Freidman, Thomas. 2005. *The World is Flat: A Brief History of the Twenty-First Century*. New York: Farrar, Straus, and Giroux.

Friedman, Thomas L. 2008. *Hot, Flat, and Crowded: Why We Need a Green Revolution and How It Can Renew America*. New York: Farrar, Straus and Giroux

Gantt, Henry Laurence. 1961. *Gantt on Management*. New York: American Management Association.

Gilbreth, Frank B. 1908. *Concrete System*. New York: Engineering News Co.

Gilbreth, Frank B.1908 *Field System*. New York: Myron C. Clark Publishing Co

Gilbreth, Frank B. 1909. *Bricklaying System*. New York: Myron C. Clark Publishing Co.

Gilbreth, Frank B. 1911. *Motion Study*. New York: Von Nostrand.

Gilbreth, Frank B. 1912. *Primer of Scientific Management*. New York: Von Nostrand.

Gilbreth, Frank B. and Lillian M. Gilbreth. 1916. *Fatigue Study*. New York: Sturgis & Walton.

Gilbreth, Frank B. and Lillian M. Gilbreth. 1917. *Applied Motion Study*. New York: Sturgis & Walton.

Gilbreth, Frank B. and Lillian M. Gilbreth. 1920. *Motion Study for the Handicapped*. New York: Macmillan.

Gilbreth, Frank B. and Lillian M. Gilbreth. 1921. *Time and Motion Study as Fundamental Factors in Planning and Control*. Newark NJ: The Mountainside Press.

Gilbreth, Frank B., Jr. 1971. *Time Out for Happiness*. New York: Crowell.

Gilbreth, Frank B., Jr. and Ernestine Carey. 1949. *Cheaper by the Dozen*. New York: Crowell.

Gilbreth, Frank B., Jr. and Ernestine Carey. 1950. *Belles on Their Toes*. New York: Crowell.

Gilbreth, Lillian M. 1914. *The Psychology of Management*. New York: Sturgis and Walton..

Gilbreth, Lillian M. 1915. "Educating the Workers for Higher Efficiency," Iron Age, 96 (December 30): 1530-33.

Gilbreth, Lillian M. 1927. *The Homemaker and Her Job*. New York: Appleton.

Gilbreth, Lillian M. 1928. *Living with Our Children*. New York: Norton.

Gilbreth, Lillian M. 1990. *The Quest of the One Best Way: A Sketch of the Life of Frank Bunker Gilbreth*. New York: Society of Women Engineers.

Gilbreth, Lillian M. 1998. *As I Remember*. Norcross, GA: Engineering and Management Press.

Gilbreth, Lillian M. and Alice Cook. 1947. *The Foreman in Manpower Management*. New York: McGraw-Hill.

Gilbreth, Lillian M., Orpha Mae Thomas, and Eleanor Clymer. 1954. *Management in the Home*. New York: Dood, Mead & Co.

Going, Charles B. 1911. *Principles of Industrial Engineering*. New York: McGraw-Hill.

Golden, Michael. 1910. "New Laboratories for Practical Mechanics." *Purdue Engineering Review*. West Lafayette IN: Purdue University Schools of Engineering. 5-7.

Goodeve, Charles,1948. "Operational Research," *Nature* 164 (1948): 377-384.

Goss, William F. 1905. *Bench Work in Wood; A Course of Study and Practice Designed for Use of Schools and Colleges*. Boston: Ginn and Co.

Goss, William F. 1907. *Locomotive Tests Conducted by the Engineering Laboratory of Purdue University*. New York: John Wiley and Sons.

Gotcher, J. Michael 1992. "Assisting the Handicapped: The Pioneering Efforts of Frank and Lillian Gilbreth." *Journal of Management* 1: 5-13.

Graham, Ben S. 1996. *People Come First*. Keynote address at WorkFlow Canada Conference. June 10, 1996. Toronto.

Graham, Laurel. 1998. *Managing On Her Own: Dr. Lillian Gilbreth and Women's Work in the Interwar Era*. Norcross, GA: Engineering and Management Press.

Graham, Loren R. 1993. *The Ghost of the Executed Engineer: Technology and the Fall of the Soviet Union*. Cambridge: Harvard University Press.

Grant, Eugene L. 1930. *Principles of Engineering Economy*. New York: Ronald Press.

Green, Constance M. 1956. *Eli Whitney and the Birth of American Technology*. London: Little, Brown, and Co.

Guillen, Mauro F. 2006. *The Taylorized Beauty of the Mechanical: Scientific Management and the Rise of Modernist Architecture*. Princeton: Princeton University Press.

Hannum, J. E. 1918. "Industrial Engineering at Purdue," *The Purdue Engineering Review* (May) 14: 68-75.

Hawkins, George A. 1960. *Engineering Education at Purdue University: A Message to the Alumni*. West Lafayette, IN: Purdue University Schools of Engineering.

Hawkins, George A., E. A. Walker, and J. M Pettit. 1968. *Goals of Engineering Education*. New York: American Society for Engineering Education.

Hawkins, George A. 1974. "Prof. George Hawkins: A Dean for All Seasons." *Extrapolation: a Newsletter for Purdue Engineering Alumni* (Spring Issue) 3: 1-4.

Hounshell, David A. 1984. *From the American System to Mass Production1800-1932.* Baltimore: The Johns Hopkins University Press.

Huber, George P. 1980. *Managerial Decision Making*. Boston: Pearson Scott.

Huber, George P. 2004. *The Necessary Nature of Future Firms: Attributes of Survivors in a Changing World.* Thousand Oaks, CA: Sage Publications.

Huber, George P. and William H. Glick, eds. 1993. *Organizational Change and Redesign: Ideas and Insights for Improving Performance.* Oxford: Oxford University Press.

Hughes, Thomas P. 1989. *American Genesis: A Century of Invention and Technological Enthusiasm*. New York: Viking.

Jacko, Julie A. and Andrew Sears, eds. 2003. *The Human-Computer Interaction Handbook.* Hillsdale, NY: Lawrence Erlbaum Associates.

Johnson, William. 1992. "Dr. Moshe Barash from the Technion to Purdue: The British Connection." *Advances in Manufacturing Processes and Systems: Symposium in Honor of Moshe Barash.* West Lafayette, IN: Purdue University School of Industrial Engineering (May 12).

Kirby, Maurice W. 2003. *Operational Research in War and Peace: the British Experience from the 1930s to 1970.* London: Imperial College Press.

Kirby, Richard S., ed. 1939. *Inventors and Engineers of Old New Haven.* New Haven: Yale University.

Kirkpatrick, Elwood G. 1970. *Quality Control for Managers and Engineers*. New York: Wiley.

Kirkpatrick, Elwood G. 1974. *Introductory Statistics and Probability for Engineering, Science, and Technology*. Englewood Cliffs, NJ: Prentice-Hall.

Knoll, H. B. 1963. *The Story of Purdue Engineering*. West Lafayette, IN: Purdue University Studies.

Kraft, Donald H., and Bert R. Boyce, 1991. *Operations Research for Libraries and Information Agencies.* Amsterdam: Elsevier Academic Press.

Krach, Keith. 2008. "The Seven Key Factors to Success as an Entrepreneur." *Purdue Engineering Impact Magazine* (December): 10.

Lancaster, Jane. 2004. *Making Time: Lillian Moller Gilbreth: A Life Beyond "Cheaper by the Dozen."* Boston: Northeastern University Press.

Lascoe, Orville D. 1968. *Plastics for Tooling*. West Lafayette, IN: Purdue University Press.

Lascoe, Orville D. 1973. *Machine Shop Operations and Setups*. Chicago: American Technical Society.

Lascoe, Orville D. 1988. *Handbook of Fabrication Processes*. Materials Park, OH: ASM Int.

Leemis, Lawrence M. 1995. *Reliability: Probabilistic Models and Statistical Methods*. Englewood Cliffs, NJ: Prentice Hall.

Leemis, Lawrence M. 2006. *Discrete-Event Simulation: A First Course*. Englewood Cliffs, NJ: Prentice-Hall.

Lehrer, Robert N. ed. 1957. *Work Simplification: Creative Thinking about Work Problems*. Englewood Cliffs, NJ: Prentice-Hall.

Lehrer, Robert N. 1965. *The Management of Improvement: Concepts, Organization, and Strategy*. New York: Reinhold.

Lehrer, Robert N. 1982. *Participative Productivity and Quality of Work Life*. Englewood Cliffs, NJ: Prentice-Hall.

Lehrer, Robert N. ed. 1983. *White Collar Productivity*. New York: McGraw-Hill.

Lehto, Mark R. and James R. Buck. 2008. *Introduction to Human Factors and Ergonomics for Engineers*. Oxford: Taylor and Francis.

Leimkuhler, Ferdinand. 1963. *Trucking of Radioactive Materials: Safety vs. Economy in Highway Transport*. Baltimore: Johns Hopkins Press.

Leimkuhler, Ferdinand and Michael Cooper. 1970. *Analytical Planning for University Libraries*. Berkeley: University of California Press.

Leimkuhler, Ferdinand and Philip Morse. 1979. "Exact Solution for the Bradford Distribution and Its Use in Modeling Information Data." *Operations Research* 27 1:187-198.

Leonhardt, David. 2009. "The Big Fix," *New York Times Magazine*. New York (Feb. 1): 27.

Liu, C. R. and T. C. Chang, eds. 1986. *Integrated and Intelligent Manufacturing*. New York: American Society of Mechanical Engineers.

Liu, C. R., A. Requicha, and S. Chandrasekar, eds. 1987. *Intelligent and Integrated Manufacturing Analysis and Synthesis*. New York: American Society of Mechanical Engineers.

Liu, C. R., T. C. Chang and R. Komanduri. 1985. *Computer-Aided/Intelligent Process Planning*. New York: American Society of Mechanical Engineers.

Maimon, Oded, Eugene Khmelnitsky, and Konstantin Kogan. 1998. *Optimal Flow Control in Manufacturing Systems: Production Planning and Scheduling*. Boston: Kluwer Academic.

Maimon, Oded and Mark Last. 2001. *Knowledge Discovery and Data Mining: The Info-fuzzy Network (IFN) Methodology.* Boston: Kluwer Academic.

Maimon, Oded and Lior Rokach. 2005. *Decomposition Methodology for Knowledge Discovery and Data Mining: Theory and Applications.* Hackensack, NJ: World Scientific.

Maimon, Oded and Lior Rokach. 2008. *Data Mining with Decision Trees: Theory and Applications.* Hackensack, NJ: World Scientific.

May, Matthew E. 2007. *The Elegant Solution: Toyota's Formula for Mastering Innovation.* Washington, DC: Free Press.

Mize Joe H. and J. Grady Cox. 1968. *Essentials of Simulation.* Englewood Cliffs, NJ: Prentice-Hall.

Mize, Joe H., Wayne Turner and Kenneth Case. 1987. *Introduction to Industrial and Systems Engineering.* Englewood Cliffs, NJ: Prentice-Hall.

Mogensen, Allan H. 1932. *Common Sense Applied to Motion Study.* New York: Factory and Industrial Management.

Moodie, Colin L. 1982. *Production Planning, Scheduling, and Inventory Control.* Norcross GA: American Institute of Industrial Engineers.

Moodie, Colin, Reha Uzsoy, and Yuehwern Yih, eds. 1995. *Manufacturing Cells: A Systems Engineering View.* Oxford: Taylor & Francis.

Morris, William T. 1975. *Work and Your Future: Living Poorer, Working Harder.* Reston, VA: Reston Publishing Co.

Morse, Philip M. 1959. *Notes on Operations Research.* Cambridge: The MIT Press.

Morse, Philip M. 1968. *Library Effectiveness: A Systems Approach.* Cambridge: The MIT Press.

Moskowitz, Herbert, Robert L. Carraway, and Thomas L. Morin. 1985. *The Stochastic Traveling Salesman Problem Revisited.* West Lafayette, IN: Purdue University Press.

Mumford , Lewis. 1967. *The Myth of the Machine.* New York: Harcourt, Brace, and World.

Mundel, Marvin E. 1947. *Systematic Motion and Time Study.* Englewood Cliffs, NJ: Prentice Hall.

Mundel, Marvin E. 1967. *A Conceptual Framework for the Management Sciences.* New York: McGraw-Hill.

Mundel, Marvin E. 1978. *Motion and Time Study: Improving Productivity.* Englewood Cliffs, NJ: Prentice Hall.

Mundel, Marvin E. 1980. *Measuring and Enhancing the Productivity of Service and Government Organizations.* Hong Kong: Asian Productivity Organization.

Mundel, Marvin E. 1983. *Improving Productivity and Effectiveness.* Englewood Cliffs, NJ: Prentice Hall.

Mundel, Marvin E. 1987. *Measuring the Productivity of Commercial Banks.* White Plains NY: Kraus International.

Mundel, Marvin E. 1987. *Measuring Total Productivity of Manufacturing Organizations: Algorithms and PC Programs.* White Plains NY: Kraus International.

Mundel, Marvin E., and David L. Danner. 1994. *Motion and Time Study: Improving Productivity.* Englewood Cliffs, NJ: Prentice Hall.

Nadler, Gerald. 1955. *Motion and Time Study.* New York: McGraw-Hill.

Nadler, Gerald. 1957. *Work Simplification.* New York: McGraw-Hill.

Nadler, Gerald. 1970. *Work Design: A Systems Concept.* Englewood Cliffs, NJ: Irwin.

Nadler, Gerald. 1981. *The Planning and Design Approach.* New York: Wiley.

Nadler, Gerald. 1982. "The Role and Scope of Industrial Engineering," *Handbook of Industrial Engineering.* New York:John Wiley. 1.3.1-1.3.17.

Nadler, Gerald, and Shozo Hibino. 1990. *Breakthrough Thinking.* Rocklin CA: Prima Publishing.

Nadworny, Milton J. 1955. *Scientific Management and the Unions 1900-1932.* Cambridge: Harvard University Press.

National Academy of Engineering. 2004. *The Engineer of 2020: Visions of Engineering in the New Century.* Washington DC: National Academies Press.

National Academy of Engineering. 2005. *Educating the Engineer of 2020: Adapting Engineering Education to the New Century.* Washington DC: National Academies Press.

National Academy of Engineering. 2007. *Rising Above the Gathering Storm.* Washington DC: National Academies Press.

National Academy of Engineering. 2008. *Changing the Conversation: Messages for Improving Public Understanding of Engineering.* Washington DC: National Academies Press.

National Science Foundation. 2008. *Universities Report Continued Decline in Real Federal S&E R&D Funding in FY 2007.* NSF Division of Science Resource Statistics Report NSF 08-320. Arlington VA: National Science Foundation.

Nelson, Daniel, ed. 1992. *A Mental Revolution: Scientific Management Since Taylor.* Columbus OH: Ohio State University Press.

Niebel, Benjamin W. 1970. *Engineering Education at Penn State.* University Park PA: Pennsylvania State University.

Nof, Shimon Y., ed. 1986. *Robotics and Material Flow*. Amsterdam: Elsevier.

Nof, Shimon Y., ed. 1994. *Information and Collaboration Models of Integration*. Boston: Kluwer Academic.

Nof, Shimon Y., ed. 1999. *Handbook of Industrial Robotics*. New York: John Wiley and Sons.

Nof, Shimon. Y., C. L. Moodie, eds. 1989. *NATO Advanced Research Workshop on Advanced Information Technologies for Industrial Material Flow Systems*. Berlin: Springer-Verlag.

Nof, Shimon Y., Wilbert E. Wilhelm, and Hans-Jürgen Warnecke. 1997. *Industrial Assembly*. Boston: Kluwer-Chapman & Hall.

Nye, David E. 2006. *Technology Matters: Questions to Live With*. Cambridge MA: The MIT Press

Ohno, Taichi. 1988. *Toyota Production System: Beyond Large Scale Production*. New York: Productivity Press.

Ovacik, I. M., and R. Uzsoy. 1997. *Decomposition Methods for Complex Factory Scheduling Problems*. Boston: Kluwer Academic.

Owen, Robert. 1821. *Report to the County of Lanark*. Glasgow: Wardlaw & Cunninghame.

Owen, Robert. 1971. *Owen: the Life of Robert Owen Written by Himself*. New York: August M. Kelley

Owen, Robert Dale. 1967. *Threading My Way: An Autobiography*. New York: August M. Kelley.

Pekny, Joseph. 2007. "Beyond Enterprise," *Purdue Engineering Impact*, (Winter): 19.

Pekny, Joseph F., Gary E. Blau, and Brice Carnahan, ed. 1998. *Foundations of Computer Aided Process Operations*. New York: Amer. Institute of Chemical Engineers.

Perkins, Robert 1895. "A Marvel of Mechanical Achievement." *The Engineering Magazine* IX: 281-301.

Pegden C. Dennis, Robert E. Shannon, Randall P. Sadowski. 1990. *Introduction to Simulation Using SIMAN*. New York: McGraw-Hill.

Pigage, Leo Charles, and J. L. Tucker. 1955. *Job Evaluation*. Urbana IL: University of Illinois Press.

Podmore, Frank. 1906. *Robert Owen: A Biography*. London: Hutchinson & Co.

Price, Brian C. 1987. *One Best Way: Frank and Lillian Gilbreth's Transformation of Scientific Management, 1885-1940*. West Lafayette, IN. Purdue University Dissertation.

Pritsker, A. Alan B. 1977. *Modeling and Analysis Using Q-GERT Networks*. New York: John Wiley.

Pritsker, A. Alan B. 1990. *Papers, Experiences, Perspectives*. West Lafayette, IN: Systems Publishing.

Pritsker, A. Alan B., 1993, *Simulation Handout*, West Lafayette, IN: Purdue University School of Industrial Engineering (unpublished).

Pritsker, A. Alan B. 1998. "Organ Transplantation Allocation Policy Analysis." *ORMS Today* 4: 22-28.

Pritsker, A. Alan B. and Steven D. Ducket. 1987. "Simulating Production Systems," *Production Handbook*, John White, ed. New York: John Wiley. 8.57-8.73.

Pritsker, A. Alan B. and Philip J. Kiviat. 1969. *Simulation with GASP II: a Fortran-Based Simulation Language*. New York: Prentice-Hall.

Pritsker, A. Alan B. and Jean J. Reilly. 1999. *Simulation with Visual SLAM and AWESIM*. New York: John Wiley and Sons.

Pritsker, A. Alan B. and C. Elliott Sigal. 1983. *Management Decision Making: A Network Simulation Approach*. Englewood Cliffs, NJ: Prentice Hall.

Purdue University. 1880. *The Sixth Annual Register 1879-80*. West Lafayette, IN.

Purdue University. 1915. *The Fortieth Annual Catalog 1914-15*. West Lafayette, IN.

Purdue University. 1928. *The Fifty-Third Annual Catalog 1927-28*. West Lafayette IN.

Purdue University. 1935. *The Sixtieth Annual Catalog 1934-35*. West Lafayette, IN.

Purdue University. 1939. *The Sixty-Fourth Annual Catalog 1938-39*. West Lafayette, IN.

Purdue University. 2008. *The One Hundred and Thirty-Third Annual Catalog 2007-2008*. West Lafayette, IN.

"Purdue's Rocket Man." 1944. *Time Magazine*. (September): 52.

Pursell, Carroll W. 1990. *Technology in America*. Cambridge: MIT Press.

Rardin, Ronald. 1998. *Optimization in Operations Research*. Englewood Cliffs, NJ: Prentice Hall.

Ravindran A., ed. 2008. *Operations Research and Management Science Handbook*. Boca Raton FL: CRC Press.

Ravindran, A., Don T. Phillips, and James J. Solberg. 1987. *Operations Research: Principles and Practice*. New York: John Wiley and Sons.

Ravindran A, K. M. Ragsdell, and G. V. Reklaitis. 2006. *Engineering Optimization: Methods and Applications*. New York: Wiley.

Reed, Ruddell. 1961. *Plant Layout: Factors, Principles, and Techniques*. Homewood IL: Irwin.

Reed, Ruddell. 1967. *Plant Location, Layout, and Maintenance*. Homewood IL: Irwin.

Reich, Charles. 1970. *The Greening of America*. New York: Random House.

Reich, Robert. 1983. *The Next American Frontier*. New York: Times Books.

Richardson, Wallace J. and Robert E. Heiland. 1957. *Work Sampling*. New York: McGraw-Hill.

Ritchey, John A. 1964. *Classics in Industrial Engineering*. West Lafayette, IN: Prairie Press.

Roberts, Stephen, Jerry Banks, and Bruce Schmeiser, eds. 1983. *Proceedings of the 1983 Winter Simulation Conference*. New York: Institute of Electrical and Electronics Engineers.

Roe, Joseph Wickham. 1916. *English and American Tool Builders*. New York: McGraw-Hill.

Roy, Robert H. 1970. *The Cultures of Management*. Baltimore MD: The Johns Hopkins University Press.

Rozak, Theodore. 1972. *Where the Wasteland Ends*. New York: Doubleday.

Runkle, John D. 1882. *The Manual Element in Education*. Boston: Rand, Avery, & Co.

Salvendy, Gavriel, ed. 1987. *Cognitive Engineering in the Design of Human-Computer Interaction and Expert Systems*. Amsterdam: Elsevier.

Salvendy, Gavriel, ed. 1997. *Handbook of Human Factors*. New York: John Wiley.

Salvendy, Gavriel, ed. 2001. *Handbook of Industrial Engineering*. New York: John Wiley.

Salvendy, Gavriel, ed. 2006. *Handbook of Human Factors and Ergonomics*. 3rd. ed. New York: John Wiley.

Salvendy, Gavriel, ed. 2007. *Human Interface and the Management of Information*. New York: Springer.

Salvendy, Gavriel, Michael J. Smith, eds. 1989. *Designing and Using Human-Computer Interfaces and Knowledge-Based Systems*. Amsterdam: Elsevier.

Salvendy, Gavriel, Michael J. Smith, Richard J. Koubek, eds. 1997. *Design of Computing Systems*. Amsterdam: Elsevier.

Samuel, Raphael E. 1977. "Workshop of the World: Steam Power and Hand Technology in Mid-Victorian Britain." *Oxford History Workshop Journal* (Spring) 3: 6-72.

Sarabia, Jose M. and Maria Sarabia. 2008. "Explicit Expressions for the Leimkuher Curve in Parametric Families." *J. of Information Processing and Management* 44: 1808-1818.

Schmeiser, Bruce W. and Reha Uzsoy, eds. 1995. *Proceedings of the Fourth Industrial Engineering Research Conference*. Norcross, GA: Institute of Industrial Engineers.

Sennett, Richard. 2008. *The Craftsman*. New Haven: Yale University Press.

Shepard, George Hugh. 1928. *The Elements of Industrial Engineering*. New York: Ginn.

Shingo, Shigeo. 1987. *The Sayings of Shigeo Shingo: Key Strategies for Plant Production*. New York: Productivity Press.

Simon, Herbert A. 1960. *The New Science of Management Decisions*. New York: Harper.

Simon, Herbert A. 1965. *The Shape of Automation for Men and Management*. New York: Harper.

Simon, Herbert A. 1981. *The Sciences of the Artificial*. Cambridge: The MIT Press.

Smith, Michael J. and Gavriel Salvendy. 1993. *Human-Computer Interaction*. Amsterdam: North-Holland.

Solberg, James J. 1989, "Production Planning and Scheduling in CIM," *Information Processing* 89: 919-925.

Solberg, James J. 2005. *Stochastic Modeling for Industrial Engineers*. Unpublished manuscript.

Solberg, James J. 2009. *Modeling Random Processes for Engineers and Managers*. Chichester, UK: John Wiley.

Sparrow, F. T. 1980. *The Iron and Steel Industry Process Model*. Washington DC: U.S. Dept. of Energy.

Sparrow, F. T., J. D. Fricker, and R. K. Whitford. 1982. *The Mobility Enterprise: Improving Auto Productivity*. West Lafayette, IN: Purdue Institute for Interdisciplinary Engineering Studies.

Spender, J. C. and Hugo J. Kijne. 1996. *Scientific Management: Frederick Winslow Taylor's Gift to the World*. Boston: Kluwer Academic Publishers.

Stecke, Kathryn E. and Rajan Suri, eds. 1985. *Flexible Manufacturing Systems: Operations Research Models and Applications*. Basel, Switzerland: J. C. Baltzer AG.

Stern, Nancy. 1979. "In the Beginning, the ENIAC." *Datamation* (May): 229-234.

Stratton, Julius M. and L. H. Mannix. 2005. *Mind and Hand: The Birth of MIT*. Cambridge: MIT Press.

Surowiecki, James. 2008. "The Open Secret of Success." *The New Yorker* (May 12): 48.

Swain, James and William Biles. 1980. *Optimization and Industrial Experimentation*. New York: Wiley.

Swanson, Don R. and Abraham Bookstein. 1972. *Operations Research Implications for Libraries*. Chicago: University of Chicago Press.

Talavage, Joseph J, F. Leimkuhler, and M. Barash. 1978. *Analysis of Management Policy for Industrial Information Centers*. West Lafayette, IN: Purdue Engineering Research Center.

Tanchoco, J. M. A., ed. 1994. *Material Flow Systems in Manufacturing.* Boston: Chapman-Hall.

Taylor, Frederick Winslow. 1947. *Scientific Management; Comprising Shop Management, The Principles of Scientific Management, and Testimony Before the Special House Committee.* New York: Harper.

Thomas, Marlin. 2006. *Reliability and Warranties: Methods for Product Development and Quality Improvement.* Boca Raton: CRC/Taylor & Francis.

Thompson, Clarence Bertrand. 1914. *Scientific Management: A Collection of the More Significant Articles Describing the Taylor System of Management.* Cambridge: Harvard University Press.

Thompson, Clarence B. 1917. *The Theory and Practice of Scientific Management.* New York: Houghton Mifflin.

Towne, Henry. 1886. "The Engineer as an Economist." *Proceedings of the ASME.* New York: ASME. 69-82.

Toynbee, Arnold. 1956. *The Industrial Revolution.* Boston: Beacon.

Tsutsui, William M. 1998. *Manufacturing Ideology: Scientific Management in Twentieth Century Japan.* Princeton: Princeton University Press.

Upton, David M., Robert H. Hayes, and Gary P. Pisano. 1996. *Strategic Operations: Competing through Capabilities.* New York: Free Press.

Urwick, Lyndall (ed.) 1956 *The Golden Book of Management.* London: Newman Neame Ltd.

Uzsoy, Reha, and I. M. Ovacik. 1997. *Decomposition Methods for Complex Factory Scheduling Problems.* Boston: Kluwer.

Van Riper, Paul. 1995. "Luther Gulick on Frederick Taylor and Scientific Management, *Journal of Management History* 2: 6-7.

Veblen, Thorstein. 1963. *The Engineers and the Price System.* New York: Harcourt, Brace & World.

Wallace, Lawrence W. 1927. "Management's Part in Maintaining Prosperity." *Bulletin of Purdue University* (December) 5: 3-10.

Wellington, Arthur Mellen. 1887. Economic Theory of the Location of Railways, New York: John Wiley and Sons.

White, John A., ed. 1987. *Production Handbook.* New York: John Wiley and Sons.

Whitford, R. K. and F. T, Sparrow. 1984. *Elderly & Handicapped Transportation Service.* West Lafayette, IN: Purdue Institute for Interdisciplinary Engineering Research Center.

Wiener, Norbert. 1954. *The Human Use of Human Beings: Cybernetics and Society* New York: Houghton Mifflin.

Wrege, Charles D. and R. G. Greenwood. 1991. *Frederick W. Taylor, the Father of Scientific Management: Myth and Reality.* Homewood, IL: Irwin.

Wysk, Richard A., Benjamin Neibel and Alan Draper. 1989. *Modern Manufacturing Process Engineering*. New York: McGraw-Hill.

Wysk, Richard A., Javier Santos and Jose M. Torres. 2006. *Improving Production with Lean Thinking*. New York: John Wiley.

Yost, Edna. 1949. *Frank and Lillian Gilbreth: Partners for Life*. New York: ASME.

Yost, Edna, and Lillian Gilbreth. 1945. *Normal Lives for the Disabled*. New York: Macmillan.

Young, Hewitt H., Gordon B. Carson and Harold A. Bolz, eds. 1972. *Production Handbook*. New York: Ronald Press.

Index

Taylor, Frederick, 3, 4, 5, 60, 61, 62, 63, 64, 65, 66, 67, 68, 72, 75, 76, 78, 82, 89, 90, 91, 93, 98, 104, 105, 107, 108, 118, 119, 121, 138, 144, 145, 157, 158, 159, 160, 161, 163, 167, 170, 184, 186, 201, 209
Thomas, Marlin, 111, 115, 123, 206
Tilles, Seymour, 111
Tompkins, James, 206, 207, 218
Towne, Henry, 89, 90, 104
Toynbee, Arnold, 29
Tseng, Mitchell M., 222
Tu, Jay, 111, 115, 143, 252
Turner, William, 39, 40, 54, 56, 97, 111, 112, 140

U

Uhan, Nelson, 111, 115, 123
Uzsoy, Reha, 111, 115, 190, 193, 206, 267

V

Veblen, Thorstein, 72, 95
Vellinger, Tony, 97, 111, 140
Venkatesan, Ravi, 220

W

Wagner, Donald, 111, 115, 123, 124, 242
Wallace, Lawrence, 91, 92, 109, 111, 112, 159, 184, 201
Walters, Jack, 111
Wan, Hong, 111, 115, 123
Wang, Peter K., 223
Watt, James, 2, 23, 31
Webster, Dennis, 206
Weinstein, Jeremy, 206, 207, 218, 246
Weldon, Thomas D., 221
White, Emerson, 3, 27, 35, 40, 54
White, Maunsel, 138
Whitney, Eli, 2, 19, 24, 30, 32, 33, 136, 137
Wiener, Norbert, 137
Wilson, James, 111, 115, 123, 206, 241, 263
Wortman, David, 190, 206, 220
Wysk, Richard, 206, 220

Y

Yang, Henry, 142
Yi, Ji Soo, 111, 115, 166, 193
Yih, Yuehwern, 111, 115, 190, 193, 267
Young, Hewitt, 101, 111, 115, 206, 260

Z

Zakaria, Adel, 207, 219, 248